The Encyclopaedic Dictionary in the Eighteenth Century
Architecture, Arts and Crafts

Volume One

John Harris

Lexicon Technicum

The Encyclopaedic Dictionary in the Eighteenth Century

Architecture, Arts and Crafts

Volume One

John Harris
Lexicon Technicum

Incorporating works of
Sir Francis Bacon and Sir Henry Wotton

Terence M. Russell

Ashgate

Published by

Ashgate Publishing Limited
Gower House
Croft Road
Aldershot
Hants GU11 3HR
England

Ashgate Publishing Limited
Old Post Road
Brookfield
Vermont 05036-9704
USA

British Library Cataloguing in Publication Data
Russell, Terence M.
The encyclopaedic dictionary in the eighteenth century: architecture, arts and crafts
Vol. 1: John Harris Lexicon Technicum: incorporating works of Sir Francis Bacon and Sir Henry Wotton
1. Architecture – Terminology – History and criticism
2. Encyclopedias and dictionaries – History and criticism
3. English language – Lexicography – History – 18th century
I. Title II. Harris, John, 1756-1846. Lexicon Technicum
720.1'4

Library of Congress Cataloging-in-Publication Data
Russell, Terence M.
The encyclopaedic dictionary in the eighteenth century: architecture, arts and crafts / Terence M. Russell.
Includes bibliographical references and indexes.
Contents: v. 1. John Harris and the Lexicon Technicum: incorporating works of Sir Francis Bacon and Sir Henry Wotton – v. 2. Ephraim Chambers Cyclopaedia – v. 3. The builders dictionary – v. 4. Samuel Johnson: a dictionary of the English language – v. 5. A society of gentlemen: the Encyclopaedia Britannica.
1. Architecture. 2. Building. I. Title
NA2540.R88 1997
720'.941'09033–DC21 96-54482
CIP

Printed on acid-free paper

ISBN 1 85928 062 5

Printed in Great Britain at the University Press, Cambridge

To

The Central Technical School Sheffield
1954-1959
In recognition of the formative influence it exerted upon me.

TMR

Johannes Harris. D. D
Regiæ Societatis Socius. Ætat. Suæ 40.

Frontispiece

John Harris at the age of thirty seven as portrayed in his cleric's robes in the painting by R. White as engraved by the illustrator G. White.

Reproduced with permission from: John Harris, *Lexicon Technicum* (1704) in Aberdeen University Library.

Contents

Illustrations

The following plates are incorporated at the end of the text:

Acknowledgements

This work has its origins in my companion publication *Architecture in the Encyclopédie of Diderot and d'Alembert: The Letterpress Articles and Selected Engravings*. On completion of this book I discussed with the publishers, Ashgate Publishing Company, the possibility of further exploring eighteenth-century writings on architectural subjects as manifest in other encyclopaedic-dictionaries of the period. Initially, the proposed project was to be a survey of the *Lexicon Technicum* of John Harris and the *Cyclopaedia* of Ephraim Chambers but the concept was eventually enlarged to incorporate *The Builder's Dictionary*, *A Dictionary of the English Language* of Samuel Johnson and *Encyclopaedia Britannica*. The result is that within the compass of a single work, albeit in multiple-volume format, we are now able to offer the reader selected writings on many facets of architecture and allied arts and crafts. I wish to express my gratitude to Mr Nigel Farrow, Chairman of the Ashgate Publishing Company, for endorsing my enthusiasm and for giving his consent for this large enterprise to proceed with the support of Ashgate Publishing. Ms Rachel Lynch, Editor at Scolar Press within the Ashgate Publishing Group, has supervised the progress of the work from its inception with diligence and unfailing courtesy and has frequently helped me to solve various procedural problems.

My orientation to many of the themes discussed in these volumes has its origins in my secondary education at a technical school and for this reason I have given expression to my debt of gratitude by inscribing each part of the work with an appropriate dedication. The scholarly impulse to the subjects here presented was nurtured by my professional training in architecture which served also to kindle my interest in the science of reference documentation and architectural bibliographical compilation. This disposition has exerted a considerable influence on the gestation and creation of my new study and for the benefit of the reader with an interest in works of reference, I have taken the liberty of listing details of my other published works in the field of architectural and historical bibliography – see *References*.

I have compiled my text from the first editions, or authoritative facsimile reproductions, of the original works which form the subject of the various volumes in the series. The researcher is well served in Scotland by a number of very fine archival collections, and I am indebted particularly to the following for their permission to cite edited passages from the original works: The National Library of Scotland, The University of Aberdeen, The University of Edinburgh and The University of St Andrews. In the case of each institution, the staff of 'Special

Collections' have been most helpful in providing me with literary treasures from their fire-proof vaults.

I am grateful to Mr Peter Hood, Convenor of the Technical Panel in Scotland of the Society for the Protection of Ancient Buildings, for allowing me to consult and to cite edited passages from his splendid copy of *The Builder's Dictionary*.

I am indebted to the trustees of the Earl of Moray Endowment Fund at the University of Edinburgh for financial assistance towards the creation of the archive of writings from which my text has been compiled. The conversion of the mass of accumulated papers into a coherent text has been a difficult undertaking that has involved the manipulation of hundreds of paper slips bearing the definitions of terms and edited abstracts of the writings selected for publication. This has imposed the need for considerable vigilance and accuracy in endeavouring to preserve the orthography, style and idiom of the original texts. In this context, I wish to record my appreciation of the efforts of Rosalind M. Russell for her contribution to the early stages of the work in the form of establishing several guiding principles and for typing an extensive trial section of the manuscript. For continuing this work and for typing the entire text and preparing it for publication with unswerving scholarly assiduity, I am immensely grateful to Thomas G. Russell.

A number of my colleagues have read sections of the manuscript and have contributed welcome suggestions, for all of which I herewith express my appreciation.

TMR

Preface

The Encyclopaedic Dictionary in the Eighteenth Century: Architecture, Arts and Crafts is the thematic title of the series of publications of which this book forms a part. The individual titles of the volumes in the series are as follows:

John Harris *Lexicon Technicum*:
Incorporating works of Sir Francis Bacon and Sir Henry Wotton
Ephraim Chambers *Cyclopaedia*
The Builder's Dictionary
Samuel Johnson and *A Dictionary of the English Language*
Encyclopaedia Britannica

The principal theme of the series is the manner in which architecture and allied arts and crafts are portrayed in the major British encyclopaedic-dictionaries of the eighteenth century. Before describing the aims of this work in greater detail it will assist the reader to say a few words first about our previous researches which have led directly to the writing of this series of encyclopaedic-dictionary studies.

The present work follows on from and is complimentary to our study *Architecture in the Encyclopédie of Diderot and d'Alembert: The Letterpress Articles and Selected Engravings*. Scolar Press, Aldershot 1993. Our new work is also a development of the author's earlier bibliographical researches, namely: *BIBLIOGRAPHICAL INDEX: A Subject Guide to Printed Books in the Fields of Architecture, Building, Constructional Engineering, Environment, Technology, Planning, Landscape Architecture, Design and allied Arts and Crafts.* University Microfilms International (UMI) Ann Arbor, Michigan USA 1982; and, *THE BUILT ENVIRONMENT: A Subject Index 1800-1960.* Gregg Publishing, London 1988.

For the reader interested in these works, additional details are provided under the heading *References*.

From the preceding information it will be seen that our new study *The Encyclopaedic Dictionary in the Eighteenth Century: Architecture, Arts and Crafts* is the consequence of a long period of gestation, in point of fact a timescale extending over more than twenty years. The primary intention of our two studies *Bibliographical Index* and *The Built Environment* has been to create a bibliographical structure for the considerable literature of what in recent years has come to be called the *built environment* or the man-made context within which we live our lives. In these two works we estimate to have collated about 75,000 bibliographical sources, bearing upon the systematic description of books with regard to their authorship, printing, publication and editions. Following on from these bibliographical researches, the opportunity arose to extend our programme of work, by undertaking a more focused investigation of historically significant encyclopaedic dictionaries.

Having provided the reader with this contextual information, we can now examine in greater detail the aims of the present work.

The principal theme of our series is the portrayal of architecture and allied arts and crafts in the major British encyclopaedic-dictionaries of the eighteenth century. The eighteenth-century impulse to compile encyclopaedic works of reference, the purpose of which was to explain the achievements of mankind to a wide readership, is well exemplified in the intellectual aspirations of the polymath and systematic thinker Wilhelm Gottfried Leibniz who conceived a project 'to extract the quintessence of the best books, to add to them the unwritten observations of the most tried in every profession, and in this manner to build systems of knowledge, based upon experience and demonstration for the further progress of mankind'. The spirit of these intentions is also captured in the views of the philosopher and statesman Sir Francis Bacon who, a generation earlier, had affirmed the value of what he considered could be gained to both philosophy and art through an increase of mutual understanding. 'Philosophy', Bacon argued, 'would be improv'd by having the Secrets of all Trades Lye open not only because such Experimental Philosophy is coucht amongst them; but also that the trades themselves might by a Philosopher be Improv'd.' Bacon's wider intellectual theories, particularly concerning the ordering of knowledge and the division of the sciences into different branches, greatly impressed the French *philosphes* Denis Diderot and Jean le Rond d'Alembert: indeed, they respectfully refer to Bacon as 'the immortal Chancellor of England'. They embodied his views in their celebrated *Preliminary Discourse* to the *Encyclopédie* in which they argued the Age-of-Enlightenment case for the advancement of learning. We consider these philosophical, intellectual and historical aspects which surround our study in greater detail in the context of the introductory chapters to the separate volumes in the series.

In our survey of the selected texts we look at the aims of each work, we consider its structure, we survey its system of arrangement and we provide a comprehensive inventory of all the writings (letterpress articles) on architectural subjects including all the allied arts and crafts. Our compilations encompass architectural terms, building construction, building materials, cabinet work, carpentry, the classical orders, decorative art, design principles including aesthetics, domes, drainage, fortification, furnishings, gardening, joinery, landscape terms, military art, mouldings, ornament, painting, perspective, sanitation, sculpture, surveying, terminology of the building trades and crafts, and the tools and equipment used in building construction.

Each article is presented under its headword and original spellings and archaisms have been retained within the orthography of the eighteenth century. Articles of all kinds have been incorporated including short, definition-like texts alongside more extended treatises. In this context particular mention may be made of our inclusion of Sir Francis Bacon's Civil Essays, *Of Building* and *Of Gardens* and Sir Henry Wotton's treatise *The Elements of Architecture* with a commentary in the form of an adaptation of Isaac Walton's celebrated *The Life of Sir Henry Wotton*.

The volumes in the series are thematically linked but each has been conceived so as to be independent of the rest with the minimum of cross referencing.

A noteworthy feature of each study is the incorporation of the author's original Preface. The inclusion of these texts preserves for the reader the guiding principles by means of which the editor-compilers created their great works of reference bearing upon such considerations as their aims and intentions; philosophy and outlook; the range and scope of their work; working method; and the many contextual considerations determined by time and place.

We believe that the writings we have extracted from *Lexicon Technicum*, *Cyclopaedia*, *The Builder's Dictionary*, *A Dictionary of the English Language* and *Encyclopaedia Britannica* provide the reader with a wealth of information about eighteenth-century building-trade practices and procedures, a conspectus of contemporary knowledge of the period, a survey of terminology and nomenclature, and a systematically presented series of theoretical and practical writings about numerous facets of architecture, building and the attendant decorative arts and crafts. The following inventory will give the reader an idea of the scope of our investigations. The number of letterpress articles extracted from their original sources is as indicated in brackets: *Harris*, Vol. I (361), Vol. II (212); *Chambers*, Vol. I (371), Vol. II (453); *Builder's Dictionary*, Vol. I (214), Vol. II (254); *Johnson*, Vol. I (349), Vol. II (395); and *Encyclopaedia Britannica*, Vol. I (167), Vol. II (230), Vol. III (216) – a total of 3,222 articles. In addition, an examination of these texts will reveal to the reader that many of our articles are in the form of extended writings which consist of numerous subordinate and secondary articles on subjects related to the themes of the principal texts. Consequently, such entries as, for example, *architecture*, *building*, *column*, *fortification*, *mosaic*, *order* and *scenography* greatly extend, through their diversity, the overall total of actual articles.

It is our hope that the volumes in our series *The Encyclopaedic Dictionary in the Eighteenth Century: Architecture, Arts and Crafts* will be of benefit to architects, artists and craftsmen, designers, restorers, preservationists, conservationists, librarians, historians, antiquarians and students. Furthermore, by incorporating within the scope of our work a systematic survey of Samuel Johnson's celebrated *A Dictionary of the English Language*, we trust our study will also be of value to the reader interested in the development of the English language bearing upon Johnson's lexicographic methodology and his definitions and descriptions of architectural and technical terms. More generally, we hope also that the books in our series, the subject matter of which ranges across architecture, history, biography and philosophy, will be regarded as having made a useful contribution to eighteenth-century, inter-disciplinary studies within the framework of the scholarship of the European Enlightenment.

It remains to take leave of the reader with a brief reference to our working method. Our first step was to individually consult each of the several hundred thousand articles embracing all manner of subjects contained within *Lexicon Technicum*, *Cyclopaedia*, The *Builder's Dictionary*, *A Dictionary of the English Language* and *Encyclopaedia Britannica* – a considerable undertaking necessitating a line-by-line scrutiny of each page of the original works from which our studies have been derived. From this vast horde of writings, inventories were compiled consisting of the headwords of all the texts relating to architecture and allied arts and crafts contained within each encyclopaedic-dictionary: these inventories are preserved in the body of our completed text in order to serve the reader as an index of contents. Working systematically through these lists, an abstract was then made on to paper slips of each text selected for inclusion in the final work. This activity, which required a number of years to bring to completion, necessitated the constant application of rigorous editorial judgement bearing upon the retention of essential information and the exclusion of extraneous material. In making our final compilations, we have erred on the side of caution in order to give the reader the maximum possible information within the scope of the limits set by the publisher. We have endeavoured throughout to be true and accurate to the original texts and to faithfully preserve the original eighteenth-century idiom If we have been negligent in places concerning such matters, we ask the reader to bear in mind the eloquent words of Theodore Besterman, doyen of editorial compilers:

> May the noble minded scholars instead of cherishing ill feeling kindly correct whatever errors have been committed . . . in the way of wrong interpretations and misstatements.

Terence M. Russell
Edinburgh, 1996.

Introduction

The first part of our study is concerned with the portrayal of architecture and allied arts and crafts in the *Lexicon Technicum* of John Harris and the representation of selected writings also on architectural subjects from the works of Sir Francis Bacon and Sir Henry Wotton.

We commence with two of Bacon's 'Civil Essays', *Of Building* and *Of Gardens*. These essays make a fitting introduction to many of the themes that we have included in our survey bearing upon the manner in which architecture and allied arts and crafts are portrayed in eighteenth-century dictionaries and encyclopaedias. For Bacon the realm of knowledge was a labyrinth that the human mind must enter, observe and from which information had to be collected as a prerequisite to understanding. By analysing experience in this inductive manner, Bacon maintained that facts and causes inherently concealed from the senses could be understood and that true conclusions could be established. In the essay *Of Building*, we have Bacon at his most practical, stating the case for the site of a building to be well considered with regard to the amenities of light, air and sunshine. From these fundamentals he proceeds to a wider discussion of room planning and the role of the decorative arts.

Next we present Bacon's celebrated essay *Of Gardens* which contains the famous opening line: 'God Almighty first planted a Garden.' Bacon was well qualified to write about gardens: in about 1608 he set about creating a series of elaborate water gardens in the grounds of his own home at Gorhambury in Hertfordshire. Bacon's views about gardens are on a princely scale: 'The contents ought not well to be under thirty acres of ground.' Notwithstanding the grandeur of his conceptions, Bacon makes many observations concerning trees, flowers and shrubs and enunciates views about fountains and principles of good landscape design that will delight and inform the reader. Moreover, Bacon's highly evocative essay provides an ideal anticipation of the writings on garden subjects which we have incorporated in our series.

We follow Bacon's writings with Sir Henry Wotton's treatise *The Elements of Architecture*. This text makes an effective bridge between the selected writings of Sir Francis Bacon and the principal theme in this part of our study, the *Lexicon Technicum* of John Harris. By way of an introduction to Wotton's famous work we have incorporated a commentary consisting of an adaptation of Isaac Walton's *The Life of Sir Henry Wotton*. Wotton and Walton shared a love of angling with their mutual friend the poet John Milton. They fished the Thames at Eton College on the bend in the river now overlooked by the Fellows' garden: the quietude and tranquillity of this location the author can vouchsafe from personal experience to be of the most idyllic of its kind to be found in the whole of England.

Walton captures the essence of Wotton's life to perfection and we believe the reader, not familiar with Walton's original biography, will find much of interest, not least the circumstances surrounding the gestation and writing of *The Elements of Architecture*.

As we have remarked, the principal theme of this volume is the *Lexicon Technicum* of John Harris. The clergyman-scholar John Harris was well equipped to undertake the intellectually demanding task of creating an encyclopaedic dictionary: he possessed a first-class intellect with a disposition to science and mathematics, he was one of the first interlocutors of Newton's epochal theories, and, perhaps most significantly, he was a punctilious editor-compiler. *Lexicon Technicum* was a pioneering work. It has the prestige of being the first book of its kind in the English language to emphasize and systematically portray writings on science, technology and allied arts and crafts. Indeed, in his day, Harris was known by the epithet *Technical* Harris. Interestingly, and erroneously, *The Oxford English Dictionary* cites Chambers' *Cyclopaedia* (1727) with the earliest recorded use of the word 'Technical'. We can now set the record straight: Harris makes clear use of the word in his Preface (1704) as follows:

> That which I have aimed at, is to make it a Dictionary not only of bare *Words* but *Things*; and that the Reader may not only find here an Explication of the *Technical* Words, or the Terms of Art made use of in all the *Liberal Sciences*, and such as border nearly upon them, but also those *Arts themselves*; and especially *such*, and *such* Parts of them as are most Useful and Advantageous to Mankind.

The use of illustrative material was rare in eighteenth-century encyclopaedias and in this regard alone, *Lexicon Technicum* was innovative. Harris, a scholar of integrity, consulted the best authors and conscientiously identified his sources. Moreover, he made it his business to derive information wherever he could at first hand. For example, in order to gain a better understanding of naval architecture, he informs us: 'I have consulted the best *Draughts*, *Sections* and *Models* I could get a sight of, have got what helps I could from Captains and Masters of Ships, and have often gone on Board my self, to get the more ready Knowledge of this Affair.'

The *Lexicon Technicum* contains numerous letterpress articles on architectural subjects which are well researched and carefully assembled. It is a remarkable one-man achievement and enjoyed a long publication run before it was overtaken by Ephraim Chambers' *Cyclopaedia*. Its creator John Harris was alert to advances in technology and recognised the importance of its application to the arts, crafts and manufactures. It is a worthy monument to a man who was both brilliant and culpably improvident and who died, as a consequence, a complete pauper.

1 Sir Francis Bacon's Civil Essays

Of Building and *Of Gardens*

Introduction

Sir Francis Bacon's *Essays or Counsels Civil and Moral* were published between 1597 and 1625 and form an important adjunct to the larger philosophical works upon which his fame rests. Bacon himself published sets of essays in 1597, in 1612, and in 1625 just one year before his death. A Latin edition was published in 1638. Bacon's Essays contain 'the earliest and latest fruits' of his wide-ranging observations on many subjects. The edition of 1597 contains no fewer than fifty-eight essays which encompass all aspects of the human condition bearing upon: *Truth*, *Death*, *Religion*, *Parents and Children*, *Marriage and Single Life*, *Love*, *Friendship*, *Riches*, *Ambition*, *Fortune*, *Beauty*, *Honour and Reputation*. In Essay XI, *Of Great Place*, Bacon asserts one of his fundamental beliefs, namely, that 'merit and good works is the end of man's motion'. Bacon's larger philosophical view of man's position in relation to nature and the World is contained in his *Instauratio Magna* (1620), the *Great Instauration*, which contains his celebrated text *Novum Organum*, *The New Organon, or, True Direction Concerning The Interpretation of Nature*. This text addresses what today we would term scientific method. Bacon argued the case for a new science and for a proper place to be given in human affairs to a rigorous system of natural philosophy that needed to be founded upon a new plan. He wished to free mankind from what he considered to be an excessively reverential attitude towards antiquity and the attitudes of mind and the precepts of mediaeval thinkers. He protested: 'The knowledge whereof the world is now possessed especially that of nature, extendeth not to magnitude and certainty of works.' He argued for a 'fresh examination of reality in all its detail'.[1]

In the Preface to *The New Organon* Bacon outlines his procedures as follows:

> Now my method, though hard to practise, is easy to explain: and it is this. I propose to establish progressive stages of certainty. The evidence of the sense, helped and guarded by a certain process of

> correction, I retain . . . I open and lay out a new path for the mind to procede in, starting directly from simple sensuous perception.

The reference to 'sensuous perception.' in Bacon's methodology is a direct anticipation of John Locke's views as outlined in his *Essay Concerning Humane Understanding* (1690). Locke, like Bacon, had a mistrust of the concept of *innate* knowledge prevalent in the seventeenth century. To Bacon, nature presented herself in the form of large classes of phenomena and manifestations that demonstrated both simple and primitive qualities which were concealed from mankind through their inherent complexity. For Bacon, the world of knowledge was a labyrinth that the human mind must enter, observe and from which information was to be collected. The key to understanding was a new method of induction. Through analysing experience, Bacon held that facts and causes concealed from the senses could be understood and that true conclusions could be established. So far reaching were Bacon's ideas that their impetus heralded the founding of the Royal Society of London.

Bacon's *Civil Essays* reveal a mind forever 'musing, searching, revolving new notions and ideas'. His writings inspired many of his own generation and Sir Henry Wotton, whose work we quote later, as Provost of Eton College recommended Bacon to be read 'in my domestic college as an ancient author.'

Bacon's essay *Of Building*, which follows, is number forty-five in the sequence of fifty-eight that Bacon had compiled by 1625. Bacon's essay *Of Building* makes an admirable introduction to many of the themes that we have included in our survey of the manner in which architecture and allied arts and crafts are portrayed in eighteenth-century dictionaries and encyclopaedias. In his essay Bacon considers the importance to a building of the site and its prospect bearing upon the amenities of light, air and sunshine; planning is examined with respect to fitness for occasion and the fulfilment of everyday practicalities; also included is the place of works of art and craftsmanship in the adornment of cabinets, cloisters, pillars and arches.[2]

OF BUILDING.

HOUSES are built to live in, and not to look on; therefore let use be preferred before uniformity, except where both may be had. Leave the goodly fabrics of houses, for beauty only, to the enchanted palaces of the poets; who build them with small cost. He that builds a fair house upon an ill seat, committeth himself to prison. Neither do I reckon it an ill seat only where the air is unwholesome; but likewise where the air is unequal; as you shall see many fine seats set upon a knap of ground, environed with higher hills round about it, whereby the heat of the sun is pent in, and the wind gathereth as in troughs; so as you shall have, and that suddenly, as great diversity of heat and cold as if you dwelt in several places. Neither is it ill air only that maketh an ill seat, but ill ways, ill markets: and, if you will consult with Momus, ill neighbours. I speak not of many more; want of water; want of wood, shade, and shelter; want of fruitfulness and mixture of grounds of several natures; want of prospect; want of level grounds; want of places at some near distance for sports of hunting, hawking, and races; too near the sea, too remote; having the commodity of navigable rivers, or the discommodity of their overflowing; too far off from great cities, which may hinder business, or too near them, which lurcheth all provisions, and maketh every thing dear; where a man hath a great living laid together, and where he is scanted: all which, as it is impossible perhaps to find together, so it is good to know them, and think of them, that a man may take as many as he can; and if he have several dwellings, that he sort them so, that what he wanteth in the one he may find in the other. Lucullus answered Pompey well; who, when he saw his stately galleries, and rooms so large and lightsome, in one of his houses, said, *Surely an excellent place for summer, but how do you in winter*? Lucullus answered, *Why, do you not think me as wise as some fowl are, that ever change their abode towards the winter*?

To pass from the seat to the house itself; we will do as Cicero doth in the orator's art; who writes books *De Oratore*, and a book he entitles *Orator*; whereof the former delivers the precepts of the art, and the latter the perfection. We will therefore describe a princely palace, making a brief model thereof. For it is strange to see, now in Europe, such huge buildings as the Vatican and Escurial and some others be, and yet scarce a very fair room in them.

First therefore, I say you cannot have a perfect palace, except you have two several sides; a side for the banquet, as is spoken of in the book of Hester, and a side for the household; the one for feasts and triumphs, and the other for dwelling. I understand both these sides to be not only returns, but parts of the front; and to be uniform without, though severally partitioned within; and to be on both sides of a great and stately tower in the midst of the front, that, as it were, joineth them together on either hand. I would have on the side of the banquet, in front, one goodly room above stairs of some forty foot high; and under it a room for a dressing or preparing place at times of triumphs. On the other side, which is the household side, I wish it divided at the first into a hall and a chapel, (with a partition between); both of good state and bigness; and those not to go all the length, but to have at the further end a winter and a summer parlour, both fair. And under these rooms, a fair and large cellar sunk under ground; and likewise some privy kitchens, with butteries and pantries, and the like. As for the tower, I would have it two stories, of eighteen foot high a piece, above the two wings; and a goodly leads upon the top, railed with statua's interposed; and the same tower to be divided into rooms, as shall be thought fit. The stairs likewise to the upper rooms, let them be upon a fair open newel, and finely railed in with images of wood cast into a brass colour; and a very fair landing-place at the top. But this to be, if you do not point any of the lower rooms for a dining place of servants. For otherwise you shall have the servants' dinner after your own: for the steam of it will come up as in a tunnel. And so much for the front. Only I understand the height of the first stairs to be sixteen foot, which is the height of the lower room.

Beyond this front is there to be a fair court, but three sides of it, of a far lower building than the front. And in all the four corners of that court fair stair-cases, cast into turrets, on the outside, and not within the row of buildings themselves. But those towers are not to be of the height of the front, but rather proportionable to the lower building. Let the court not be paved, for that striketh up a great heat in summer, and much cold in winter. But only some side alleys, with a cross and the quarters to graze, being kept shorn, but not too near shorn. The row of return on the banquet side, let it be all stately galleries: in which galleries let there be three, or five, fine cupolas in the length of it, placed at equal distance; and fine coloured windows of several works. On the house-hold side, chambers of presence and ordinary entertainments, with some bed-chambers; and let all three sides be a double house, without thorough lights on the sides, that you may have rooms from the sun both for forenoon and afternoon. Cast it also, that you may have rooms both for summer and winter; shady for summer, and warm for winter. You shall have sometimes fair houses so full of glass, that one cannot tell where to become to be out of the sun or cold. For in bowed windows, I hold them of good use (in cities, indeed, upright do better, in respect of the uniformity towards the street); for they be pretty retiring places for conference; and besides they keep both the wind and sun off; for that which would strike almost thorough the room doth scarce pass the window. But let them be but few, four in the court, on the sides only.

Beyond this court let there be an inward court, of the same square and height; which is to be environed with the garden on all sides; and in the inside, cloistered on all sides, upon decent and beautiful arches, as high as the first story. On the under story, towards the garden, let it be turned to a grotta, or place of shade, or estivation. And only have opening and windows towards the garden; and be level upon the floor, no whit sunken under ground, to avoid all dampishness. And let there be a fountain, or some fair work, of statua's in the midst of this court; and to be paved as the other court was. These buildings to be for privy lodgings on both sides; and the end for privy galleries. Whereof you must foresee that one of them be for an infirmary, if the prince or any special person should be sick, with chambers, bed-chamber, antecamera, and recamera, joining to it. This upon the second story. Upon the ground story, a fair gallery, open, upon pillars; and upon the third story likewise, an open gallery, upon pillars, to take the prospect and freshness of the garden. At both corners of the further side, by way of return, let there be two delicate or rich cabinets, daintily paved, richly hanged, glazed with crystalline glass, and a rich cupola in the midst; and all other elegancy that may be thought upon. In the upper gallery too, I wish that there may be, if the place will yield it, some fountains running in divers places from the wall, with some fine avoidances. And thus much for the model of the palace, save that you must have before you come to the front, three courts. A green court plain, with a wall about it; a second court of the same, but more garnished, with little turrets, or rather embellishments, upon the wall; and a third court, to make a square with the front, but not to be built, nor yet enclosed with a naked wall, but enclosed with tarrasses, leaded aloft, and fairly garnished, on the three sides; and cloistered on the inside, with pillars, and not with arches below. As for offices, let them stand at distance, with some low galleries, to pass from them to the palace itself.

The second of our extracts from the writings of Sir Francis Bacon is his essay *Of Gardens*. This text has become celebrated through its opening line 'God Almighty first planted a Garden.'[3]

Bacon was well qualified to write about gardens. In 1601 he inherited his father's house at Gorhambury in Hertfordshire. The house had been built between 1563 and 1568 by Sir Nicholas Bacon who, like his son Sir Francis, was Lord Chancellor and Lord Keeper of the Seal. The house though modest in scale for a man of Bacon's position, inspired him to realise some of his ideas concerning gardens. About 1608 he set about creating an elaborate water garden in the pondyards on the estate. Bacon's garden was square and moated and incorporated a lake with an island on which was built a large summer house complete with 'galleries, dining rooms, sleeping chambers, and private garden.'

Bacon's essay *Of Gardens* is highly evocative and makes a fitting anticipation of the writings on gardens and garden subjects which we include in subsequent sections of our survey of architectural and related subjects.[4]

Of Gardens.

GOD ALMIGHTY first planted a Garden. And indeed it is the purest of human pleasures. It is the greatest refreshment to the spirits of man; without which buildings and palaces are but gross handyworks: and a man shall ever see that when ages grow to civility and elegancy, men come to build stately sooner than to garden finely; as if gardening were the greater perfection. I do hold it, in the royal ordering of gardens, there ought to be gardens for all the months in the year; in which severally things of beauty may be then in season. For December, and January, and the latter part of November, you must take such things as are green all winter: holly; ivy; bays; juniper, cypress-trees; yew; pine-apple-trees; fir-trees; rosemary; lavender; periwinkle, the white, the purple, and the blue; germander; flags; orange-trees; lemon-trees; and myrtles, if they be stoved; and sweet marjoram, warm set. There followeth for the latter part of January and February, the mezereon-tree, which then blossoms; crocus vernus, both the yellow and the grey; primroses; anemones; the early tulippa; hyacinthus orientalis; chamaïris; fritellaria. For March, there come violets, specially the single blue, which are the earliest; the yellow daffodil; the daisy; the almond-tree in blossom; the peach-tree in blossom; the cornelian-tree in blossom; sweet-briar. In April follow, the double white violet; the wall-flower; the stock-gilliflower; the cowslip; flower-de-lices, and lilies of all natures; rosemary-flowers; the tulippa; the double-piony; the pale daffodil; the French honeysuckle; the cherry-tree blossom; the dammasin and plum-trees in blossom; the white thorn in leaf; the lilac-tree. In May and June come pinks of all sorts, specially the blush-pink; roses of all kinds, except the musk, which comes later; honey-suckles; strawberries; bugloss; columbine; the French marigold; flos Africanus; cherry-tree in fruit; ribes; figs in fruit; rasps; vine flowers; lavender in flowers; the sweet satyrian, with the white flower; herba muscaria; lilium convallium; the apple-tree in blossom. In July come gilliflowers of all varieties; musk roses; the lime-tree in blossom; early pears and plums in fruit; genitings, quadlins. In August come plums of all sorts in fruit; pears; apricocks, berberries; filberds; musk-melons; monks-hoods, of all colours. In September come grapes; apples; poppies of all colours; peaches; melocotones; nectarines; cornelians; wardens; quinces. In October and the beginning of November come services; medlars; bullaces; roses cut or removed to come late; holly-oaks; and such like. These particulars are for the climate of London; but my meaning is perceived, that you may have *ver perpetuum*, as the place affords.

And because the breath of flowers is far sweeter in the air (where it comes and goes like the warbling of music) than in the hand, therefore nothing is more fit for that delight, than to know what be the flowers and plants that do best perfume the air. Roses, damask and red, are fast flowers of their smell; so that you may walk by a whole row of them, and find nothing of their sweetness; yea though it be in a morning's dew. Bays likewise yield no smell as they grow. Rosemary little; nor sweet marjoram. That which above all others yields the sweetest smell in the air, is the violet, specially the white double violet, which comes twice a year; about the middle of April, and about Bartholomew-tide. Next to that is the musk-rose. Then the strawberry-leaves dying, with a most excellent cordial smell. Then the flower of the vines; it is a little dust, like the dust of a bent, which grows upon the cluster in the first coming forth. Then sweet-briar. Then wall-flowers, which are very delightful to be set under a parlour or lower chamber window. Then pinks and gilliflowers, specially the matted pink and clove gilliflower. Then the flowers of the lime-tree. Then the honeysuckles, so they be somewhat afar off. Of bean-flowers I speak not, because they are field flowers. But those which perfume the air most delightfully, not passed by as the rest, but being trodden upon and crushed, are three; that is, burnet, wild-thyme, and watermints. Therefore you are to set whole alleys of them, to have the pleasure when you walk or tread.

For gardens (speaking of those which are indeed prince-like, as we have done of buildings) the contents ought not well to be under thirty acres of ground; and to be divided into three parts; a green in the entrance; a heath or desert in the going forth; and the main garden in the midst; besides alleys on both sides. And I like well that four acres of ground be assigned to the green; six to the heath; four and four to either side; and twelve to the main garden. The green hath two pleasures: the one, because nothing is more pleasant to the eye than green grass kept finely shorn; the other, because it will give you a fair alley in the midst, by which you may go in front upon a stately hedge, which is to enclose the garden. But because the alley will be long, and, in great heat of the year or day, you ought not to buy the shade in the garden by going in the sun thorough the green, therefore you are, of either side the green, to plant a covert alley, upon carpenter's work, about twelve foot in height, by which you may go in shade into the garden. As for the making of knots or figures with divers coloured earths, that they may lie under the windows of the house on that side which the garden stands; they be but toys: you may see as good sights many times in tarts. The garden is best to be square, encompassed on all the four sides with a stately arched hedge. The arches to be upon pillars of carpenter's work, of some ten foot high, and six foot broad; and the spaces between of the same dimension with the breadth of the arch. Over the arches let there be an entire hedge of some four foot high, framed also upon carpenter's work; and upon the upper hedge, over every arch a little turret, with a belly, enough to receive a cage of birds: and over every space between the arches some other little figure, with broad plates of round coloured glass gilt, for the sun to play upon. But this hedge I intend to be raised upon a bank, not steep, but gently slope, of some six foot, set all with flowers. Also I understand, that this square of the garden should not be the whole breadth of the ground, but to leave on either side ground enough for diversity of side alleys; unto which the two covert alleys of the green may deliver you. But there must be no alleys with hedges at either end of this great enclosure; not at the hither end, for letting you prospect upon this fair hedge from the green; nor at the further end, for letting your prospect from the hedge through the arches upon the heath.

For the ordering of the ground within the great hedge, I leave it to variety of device; advising nevertheless that whatsoever form you cast it into, first, it be not too busy, or full of work. Wherein I, for my part, do not like images cut out in juniper or other garden stuff; they be for children. Little low hedges, round, like welts, with some pretty pyramides, I like well; and in some places, fair columns upon frames of carpenter's work. I would also have the alleys spacious and fair. You may have closer alleys upon the side grounds, but none in the main garden. I wish also, in the very middle, a fair mount, with three ascents, and alleys, enough for four to walk abreast; which I would have to be perfect circles, without any bulwarks or embossments; and the whole mount to be thirty foot high; and some fine banqueting house; with some chimneys neatly cast, and without too much glass.

For fountains, they are a great beauty and refreshment; but pools mar all and make the garden unwholesome, and full of flies and frogs. Fountains intend to be of two natures: the one that sprinkleth or spouteth water; the other a fair receipt of water, of some thirty or forty foot square, but without fish, or slime, or mud. For the first, the ornaments of images gilt, or of marble, which are in use, do well: but the main matter is so to convey the water, as it never stay, either in the bowls or in the cistern; that the water be never by rest discoloured, green or red or the like; or gather any mossiness or putrefaction. Besides that, it is to be cleansed every day by the hand. Also some steps up to it, and some fine pavement about it, doth well. As for the other kind of fountain, which we may call a bathing pool, it may admit much curiosity and beauty; wherewith we will not trouble ourselves; as, that the bottom be finely paved, and with images; the sides likewise; and withal embellished with coloured glass, and such things of lustre; encompassed also with fine rails of low statua's. But the main point is the same which we mentioned in the former kind of fountain; which is, that the water be in perpetual motion, fed by a water higher than the pool, and delivered into it by fair spouts, and then discharged away under ground, by some equality of bores, that it stay little. And for fine devices, of arching water without spilling, and making it rise in several forms (of feathers, drinking glasses, canopies, and the like), they be pretty things to look on, but nothing to health and sweetness.

For the health, which was the third part of our plot, I wish it to be framed, as much as may be, to a natural wildness. Trees I would have none in it, but some thickets made only of sweet-briar and honeysuckle, and some wild vine amongst; and the ground set with violets, strawberries, and primroses. For these are sweet, and proper in the shade. And these to be in the heath, here and there, not in any order. I like also little heaps, in the nature of mole-hills (such as are in wild heaths), to be set, some with wild thyme; some with pinks; some with germander, that gives a good flower to the eye; some with periwinkle; some with violets; some with strawberries; some with cowslips; some with daisies; some with red roses; some with lilium convallium; some with sweet-williams red; some with bear's-foot: and the like low flowers, being withal sweet and sightly. Part of which heaps are to be with standards of little bushes pricked upon their top, and part without. The standards to be roses; juniper; holly; berberries (but here and there, because of the smell of their blossom); red currants; gooseberry; rosemary; bays; sweet-briar, and such like. But these standards to be kept with cutting, that they grow not out of course.

For the side grounds, you are to fill them with variety of alleys, private, to give a full shade, some of them, wheresoever the sun be. You are to frame some of them likewise for shelter, that when the wind blows sharp, you may walk as in a gallery. And those alleys must be likewise hedged at both ends, to keep out the wind; and these closer alleys must be ever finely gravelled, and no grass, because of going wet. In many of these alleys likewise, you are to set fruit-trees of all sorts; as well upon the walls as in ranges. And this would be generally observed, that the borders wherein you plant your fruit-trees be fair and large, and low, and not steep; and set with fine flowers, but thin and sparingly, lest they deceive the trees. At the end of both the side grounds, I would have a mount of some pretty height, leaving the wall of the enclosure breast high, to look abroad into the fields.

For the main garden, I do not deny but there should be some fair alleys ranged on both sides, with fruit trees; and some pretty tufts of fruit trees, and arbours with seats, set in some decent order; but these to be by no means set too thick; but to leave the main garden so as it be not close, but the air open and free. For as for shade, I would have you rest upon the alleys of the side grounds, there to walk, if you be disposed, in the heat of the year or day; but to make account that the main garden is for the more temperate parts of the year; and in the heat of summer, for the morning and the evening, or overcast days.

For aviaries, I like them not, except they be of that largeness as they may be turfed, and have living plants and bushes set in them: that the birds may have more scope, and natural nestling, and that no foulness appear in the floor of the aviary. So I have made a platform of a princely garden, partly by precept, partly by drawing, not a model, but some general lines of it; and in this I have spared no cost. But it is nothing for great princes, that for the most part taking advice with workmen, with no less cost set their things together; and sometimes add statua's, and such things for state and magnificence, but nothing to the true pleasure of a garden.

2 Sir Henry Wotton and *The Elements of Architecture*

Sir Henry Wotton: biographical note

Sir Henry Wotton was born 30 March 1568 at Boughton Hall in the parish of Boughton, Kent. The life of Wotton was written by his friend Isaac Walton with whom Wotton shared a love of angling. The following account is drawn largely from Walton's study *The Life of Sir Henry Wotton.*[1]

Wotton's appreciation of architecture may have been nurtured in childhood amidst the agreeable surroundings of the family home:

> Bocton-hall being an ancient and goodly structure, beautifying, and being beautified by the parish-church of Bocton Malherbe [Boughton Malherbe] adjoining unto it, and both seated within a fair park of the Wottons, on the brow of such a hill as gives the advantage of a large prospect, and of equal pleasure to all beholders.

Wotton's father was Thomas Wotton who was born in 1521. From him, Wotton inherited his disposition to the arts, the character of a connoisseur and the bearing of a diplomat.

> He was a gentleman excellently educated, and studious in all the liberal arts: in the knowledge whereof he attained unto a great perfection; who, though he had (beside those abilities, a very noble and plentiful estate and the ancient interest of his predecessors) many invitations from Queen Elizabeth to change his country recreations and retirement for a court, offering him a knighthood (she was then with him at Bocton-hall), and to be but as an earnest of some more honourable and profitable employment under her; yet he humbly refused both, being a man of great modesty, of a most plain and single heart, of an ancient freedom and integrity of mind. A commendation which Sir Henry Wotton took occasion to remember with great gladness, and thankfully to boast himself the son of a father; from whom indeed he derived that noble ingenuity that was always practised by himself, and which he ever both commended and cherished in others.

Thomas Wotton had four sons, Sir Edward, Sir James, Sir John and Sir Henry. Edward was knighted by Queen Elizabeth and was made Comptroller of her Majesty's Household. James, the second son, was knighted in the field near Cádiz 'after they had gotten great honour and riches, besides a notable retaliation of injuries by taking that town [Cádiz].' John was 'a gentleman excellently accomplished, both by learning and travel' who also was knighted by Queen Elizabeth and who looked upon him 'with more than ordinary favour, and with intentions of preferment; but death in his younger years put a period to his growing hopes.' Sir Henry Wotton was the son of Thomas Wotton's second wife, the widow Mrs Elionora Morton. She was his tutoress during much of his early childhood, for whose care and pains 'he paid her each day with such visible signs of future perfection in learning, as turned her employment into a pleasing trouble.' At a later date, Thomas Wotton placed the education of his promising son under the care of a tutor until he was sent to commence his formal studies at Winchester College.

The regime at Winchester was one of strict discipline and order 'so that he might in his youth be moulded into a method of living by rule, which his wise father knew to be the most necessary way, to make the future part of his life happy to himself, and useful for the discharge of all business, whether public or private.' Wotton entered Oxford University as a commoner of New College from which he proceeded at the age of about eighteen to Queen's College. It was here he wrote a tragedy *Tancredo* (which has not survived) that impressed those who read it through the effect of its 'interwoven sentences' and for the method and exact 'personating those humours, passions, and dispositions, which he proposed to represent'. The gravest of Oxford's society considered Wotton had given 'an early and a solid testimony of his future abilities'.

In his twentieth year, Wotton graduated MA and distinguished himself by giving three lectures on optics (*de Oculo*) in which he addressed the proposition: 'Whether we see by the emission of the beams from within, or the reception of the species from without?'. His conclusion provides a fine example of the eloquence that characterises his later prose style:

> Seeing:– By which, we do not only discover Nature's secrets, but with a continued content (for the eye is never weary of seeing) behold the great light of the world, and by it discover the fabric of the heavens, and both the order and motion of the celestial orbs; nay, that if the eye look but downward, it may rejoice to behold the bosom of the earth, our common mother, embroidered and adorned with numberless and various flowers, which man sees daily grow up to perfection, and then silently moralize his own condition, who in a short time (like those very flowers) decays, and withers, and quickly returns again to that earth from which both had their first being.

Following his father's death, Wotton stayed in Oxford for about two years when he 'laid aside his books, and betook himself to the useful library of travel, and a more general conversation with mankind: employing the remaining part of his youth, his industry and fortune, to adorn his mind, and to purchase the rich treasure of foreign knowledge.' He travelled extensively in Europe between 1589 and 1594 developing his linguistic skills to a level that 'his command of German and Italian was such that he could pass as either'. It is from this early period that Wotton began to develop his understanding of architecture and painting, observing everything as he travelled with a keen eye and collecting pictures for himself and his friends: it is said that such was his judgement on artistic matters that, later in life, no less than Charles I deferred to his taste. Whilst in Italy, Wotton lived in Rome, Venice and Florence where he became acquainted with 'the most eminent men for learning, and all manner of arts; as Picture, Sculpture, Chymistry, Architecture, and other manual arts, even of inferior nature; of all which he was a most dear lover, and a most excellent judge.'

He returned to England when about thirty years of age by which time his youthful sharp wit and aptness for jesting had matured into 'a most persuasive behaviour: which was so mixed with sweet discourse and civilities, as gained him love from all persons with whom he entered into an acquaintance.' Wotton's attainments earned him a position with Robert Devereux, second Earl of Essex and ill-fated favourite of Queen Elizabeth, with whom he remained for a few years. Wotton was too sagacious to allow himself to become embroiled in Essex's rebellious indiscretions and 'thought prevention by absence out of England a better security than to stay in it, and there plead his innocency in prison.' Accordingly, Wotton left England via Dover 'without so much as looking toward his native and beloved Bocton' and 'by the help of favourable winds and liberal payment of the mariners, within sixteen hours after his departure from London, set upon the French shore; where he heard shortly after, that the Earl was arraigned, condemned, and beheaded . . . and divers other persons of eminent quality executed.'

Wotton made his way to Italy where he was taken as Secretary to the Grand Duke of Tuscany, a tribute to his intellectual and linguistic accomplishments. Whilst in Florence he compiled *The State of Christendom; or a most Exact and Curious Discovery of many Secret Passages and Hidden Mysteries of the Times*. The laudatory style of this work adopts the language of the panegyric and was probably intended by Wotton to ingratiate himself with Queen Elizabeth. Wotton's attitude at this time to his adopted home was ambivalent. He could readily identify with the view: 'That men of learning should fix their residence at Florence we need not wonder, when we reflect that this city has been long celebrated for its many excellent libraries, and principally for the Ducal palace, which contains the greatest and most valuable collection made by one family, and within one roof, of ancient and modern sculpture, paintings, and curiosities of every kind, both natural and artificial.' At the same time he could write from Florence to Lord Zouch (1592): 'I live here in a Paradise inhabited by devils. Venice hath scarce heard of these views which are here practised.'

A dramatic event occurred at this time which occasioned Wotton's hasty departure from Italy and later turned to his considerable advantage. The circumstance of Queen Elizabeth being in her decline had raised the question as to the name of her successor. James VI, King of Scotland, was confidently believed by most to be 'the man upon whom the sweet trouble of kingly government would be imposed.' For reasons of Catholic and Protestant plottings his name was not however universally acceptable and Ferdinand, the Grand Duke of Tuscany, learned of calumnies and a plot to murder James which occasioned him to entrust to Wotton letters warning James of his danger. Wotton was dispatched to Scotland with the letters and such was Ferdinand's concern he also sent Italian antidotes against poison 'as the Scots till then had been strangers to.' To effect his safe departure from Italy Wotton assumed the disguise of one Octavio Baldi, his proficiency in Italian enabling him to carry off the masquerade convincingly. Notwithstanding the strength of his disguise, Wotton took the added precaution of throwing off the suspicion that any would-be pursuer might harbour by approaching Scotland through first despatching himself via the circuitous route to Norway.

Wotton had to maintain something of a cloak-and-dagger existence for several weeks as he completed his long journey until he eventually found James with his court at Stirling. Initially the King, receiving intimation from his aids of the apparent gravity of the news borne by this strange intruder, consented only to receiving the still disguised Wotton in the presence of several suitably armed, trusty Scotch lords and only after Wotton had consented to surrender the long Italian rapier that he had been wearing for his own protection. Isaac Walton tells us:

> Then did Octavio Baldi deliver his letters and his message to the King: which when the King had graciously received, after a little pause, Octavio Baldi steps to the table, and whispers to the King in his own language, that he was an Englishman, beseeching him for a more private conference with his Majesty, and that he might be concealed during his stay in that nation; which was promised, and really performed by the King during his abode there, which was about three months: all which time was spent with much pleasantness to the King, and with as much to Octavio Baldi himself as that country could afford; from which he departed as true an Italian as he came thither.

After a stay of some months in Scotland, Wotton returned to Florence where news later reached him of the death of Queen Elizabeth and of James's accession to the English throne. Of this circumstance, Walton gives the following absorbing account:

When King James came into England, he found, amongst other of the late Queen's officers, Sir Edward [Wotton], who was after Lord Wotton, Comptroller of the House, of whom he demanded, 'If he knew one Henry Wotton, that had spent much time in foreign travel?' The Lord replied, he knew him well, and that he was his brother: Then the King, asking where he then was, answered, at Venice or Florence; but by late letters from thence he understand he would suddenly be at Paris. [We should note here that Wotton, on receiving news of James's elevation, had lost no time in seeking preferment by writing to the statesman Sir Robert Cecil offering his services.] 'Send for him,' said the King; 'and when he shall come into England, bid him repair privately to me.' The Lord Wotton, after a little wonder, asked the King, 'if he knew him?' to which the King answered, 'you must rest unsatisfied of that till you bring the gentleman to me.'

Not many months after this discourse, the Lord Wotton brought his brother to attend the King, who took him in his arms, and bade him welcome, by the name of *Octavio Baldi;* saying he was the most honest, and therefore the best dissembler that ever he met with: And said, 'Seeing I know you neither want learning, travel, nor experience, and that I have had so real a testimony of your faithfulness and abilities to manage ambassage, I have sent for you to declare my purpose; which is, to make use of you in that kind hereafter.' And indeed the King did so most of those two-and-twenty years of his reign; but before he dismissed Octavio Baldi from his present attendance upon him, he restored him his old name of Henry Wotton, by which he then knighted him.

Wotton's estate at Boughton Hall (Bocton Hall) was relatively modest by the standards of a gentleman and somewhat confining to his courtly ambitions and predilections for fine things. Not surprisingly, when Wotton was invited to accept the post of ambassador to Venice he seized the opportunity knowing full well 'the courts of great princes to be sumptuous, and necessarily expensive, inclined most to that of Venice, as being a place of more retirement, and best suiting with his genius, who did ever love to join with business, study, and a trial of natural experiments: for both which, fruitful Italy, that darling of Nature, and cherisher of all arts, is so justly famed in all parts of the Christian world.'

Wotton had calculated his chances well as this passage from Walton's account testifies:

Sir Henry having after some short time and consideration resolved upon Venice, and a large allowance being appointed by the King for his voyage thither, and a settled maintenance during his stay there, he left England [1604], nobly accompanied through France to Venice by gentlemen of the best families and breeding that this nation afforded: they were too many to name, but these two, for the following reasons may not be omitted. Sir Albertus Morton his nephew, who went his secretary; and William Beded, a man of choice learning, and sanctified wisdom, who went his chaplin.

Whilst he had been an undergraduate at Oxford, Wotton had established a lasting friendship with the poet and priest John Donne. As a mark of his affection for Wotton, Donne bid farewell to his friend on the eve of his departure to Italy with verses which include the following lines:

Admit this honest paper; and allow
It such an audience as yourself would ask;
What you would say at Venice, this says now,
And has for nature what you have for task.

'Tis therefore well your spirits now are plac'd
In their last furnace, in activity,
Which fits them: schools, and courts, and wars o'er-past,
To touch and taste in any best degree.

But though she parts us, to hear my oft prayers
For your increase, God is as near here:
And, to send you what I shall beg, his stairs
In length and ease are alike every where.

Wotton was well received in Venice with 'much honour and gladness, both for delivering his ambassge most elegantly in the Italian language, and came also in such a juncture of time, as his master's friendship seemed useful to that republic.' His long sojourn in Italy was 'fair and highly valued' until an incident clouded his reputation. It is related that Wotton, having occasion to pass through Germany, was asked by friends to give some demonstration of his learning and ingeniousness. Entering into the merriment of the occasion, Wotton composed his famous line:

An ambassador is an honest man, sent to lie abroad for the good of his country.

It is conceivable the incident would have passed without notice save for the fact that one Jasper (Casper) Scioppius, 'a Romanist, a man of a restless spirit and a malicious pen', published a book attacking James I in which he gave prominence to Wotton's levity. Furthermore, it was 'presently after written in several glass-windows, and spitefully declared to [be] Sir Henry Wotton's.' Wotton acted quickly to repair the damage inflicted upon his reputation by writing two apologies which were both 'so ingenious, so clear, and so choicely eloquent, that his Majesty (who was a pure judge of it) could not forbear, at the receipt thereof, to declare publicly, that "Sir Henry Wotton had commuted sufficiently for a greater offence".'

In his *The Life of Sir Henry Wotton*, Isaac Walton draws a parallel, which we should not overlook, with the sentiments expressed in Wotton's letters of apology and the 'useful fancy' to be found in lines from Donne's poem *The Will*:

> I give my reputation to those
> Which were my friends; mine industry to foes;
> To schoolmen I bequeath my doubtfulness;
> My sickness to physicians, or excess;
> To Nature all that I in rhyme have writ,
> And to my company my wit.

Walton concludes this hazardous episode by reflecting: 'And now, as broken bones, well set, become stronger; so Sir Henry Wotton did not only recover, but was much more confirmed in his Majesty's estimation and favour than formerly he had been.' In point of fact, Wotton's reputation took somewhat longer to repair than Walton makes out and during his temporary period of disgrace Wotton, who had good cause to reflect on his personal circumstances, composed his poem *Character of a Happy Life*.

During a period in England, Wotton was in the Parliament of 1614 as member for Appleby and later in that year he was sent on a special mission to The Hague. The following year he was re-appointed ambassador to Venice where he remained before finally returning home in 1623. Throughout his long stay overseas, Wotton gave frequent eloquent testimony to his poetical abilities, none more tellingly than in the verses he composed for the Queen of Bohemia, the opening lines of which have achieved a degree of fame eclipsed only by the celebrated opening lines of his *Elements of Architecture*. To round out our picture of Sir Henry Wotton and his many-sided accomplishments, we reproduce the lines to his 'dear mistress' below:

Elizabeth of Bohemia

You meaner beauties of the night,
That poorly satisfy our eyes
More by your number than your light,
You common people of the skies,
What are you when the sun shall rise?

You curious chanters of the wood,
That warble forth dame Nature's lays,
Thinking your voices understood
By your weak accents; what's your praise,
When Philomel her voice shall raise?

You violets that first appear,
By your pure purple mantles known,
Like the proud virgins of the year,
As if spring were all your own,
What are you when the rose is blown?

So when my mistress shall be seen,
In form and beauty of her mind,
By virtue first, then choice a Queen,
Tell me, if she were not design'd
The eclipse and glory of her kind.

Upon his finally settling in England in 1623 Wotton was in urgent need of securing an appointment that would maintain him, his servants and the several German and Italian artists he had brought with him. Providentially for Wotton a circumstance then occurred which provided him with the opportunity he was seeking. The Provostship of Eton College had become vacant on the death of the incumbent Thomas Murray. This coveted post provided an income, accommodation and a position of high social standing befitting the former King's ambassador to Venice. A measure of the prestige attaching to the appointment can be gauged from the realisation that a rival candidate for the post was no less a figure than the mighty Francis Bacon, sometime Lord Chancellor of England and Viscount of St. Albans.

So it came about that 'Sir Henry, who had for many years (like Sisyphus) rolled the restless stone of a state employment, knowing experimentally that the great blessing of sweet content was not to be found in multitude of men or business, and that a College was the fittest place to nourish holy thoughts, and to afford rest both to his body and mind, which his age (being now almost three-score years) seemed to require, did therefore use his own and the interest of all his friends to procure that place.' The effort which Wotton expended to secure the

Provostship of Eton College was indeed considerable and indirectly was the cause of him writing the subject of this part of our study, namely, *The Elements of Architecture*. We look at the circumstances under which Wotton created his celebrated text later.

Wotton was instituted into the Provostship on 26 July 1624. The post was a comfortable one having, in addition to all manner of graceful appurtenances, sufficient for 'his diet, his lodging, his horse-meat, his servants' wages, his riding charges, and 100 pounds per annum besides.'

The College statutes required Wotton to enter into Holy Orders which he did 'being made Deacon with all convenient speed.' Walton gives us the following brief insight into his friend's daily routine at this period:

> After his customary public devotions, his use was to retire into his study, and there to spend some hours in reading the Bible and authors in divinity, closing up his meditations with private prayer. This was for the most part his employment in the forenoon. But when he was once sat to dinner, then nothing but cheerful thoughts possessed his mind, and those still increased by constant company at his table of such persons as brought thither additions both of learning and pleasure; but some part of most days was usually spent in philosophical conclusions. Nor did he forget his innate pleasure of angling, which he would usually call 'His idle time not idly spent'; saying often, 'He would rather live five May months than forty Decembers.'

Wotton undertook his responsibilities as an educator seriously and was something of an innovator in the use of visual materials to assist with the instruction of his pupils as the following passage from Walton's account reveals:

> He was a constant cherisher of all those youths in the school, in whom he found either a constant diligence or a genius that prompted them to learning. For whose encouragement he was . . . at the charge of setting up in it two rows of pillars, on which he caused to be choicely drawn the pictures of divers of the most famous Greek and Latin historians, poets, and orators: persuading them not to neglect rhetoric, because 'Almighty God has left mankind affections to be wrought upon.'

Wotton filled out the years remaining to him in quiet, purposeful industry retiring once a year to his ancestral home at Boughton Hall where he would say: 'He found a cure for all cares, by the cheerful company, which he called the living furniture of the place: and a restoration of his strength, by the connaturalness of that which he called his genial air.' He was also given to visiting Oxford and his old school Winchester College which evoked fond memories of his own youth and his early education which he recalled in a reminiscence: 'And seeing that very place where I sat when I was a boy, occasioned me to remember those very thoughts of my youth which then possessed me: sweet thoughts, indeed, that promised my growing years numerous pleasures without mixtures of cares, and those to be enjoyed when time (which I therefore thought slow paced) had changed my youth into manhood.'

Wotton died in December 1639. His friend Isaac Walton concludes his *Life* as follows:

> And thus the circle of Sir Henry Wotton's life – that circle which began at Bocton, and in the circumference thereof did first touch at Winchester school, then at Oxford, and after upon so many remarkable parts and passages in Christendom – that circle of his life was by death thus closed up and completed, in the seventy-second year of his age, at Eaton College, where, according to his will, he now lies buried, with his motto on a plain grave-stone over him. Dying worthy of his name and family; worthy of the love and favour of so many Princes and persons of eminent wisdom and learning; worthy of the trust committed unto him for the service of his Prince and country.

The origins of Wotton's The Elements of Architecture

In our summary account of the life of Sir Henry Wotton we have noted that he had to make a singular effort to secure the much-needed Provostship of Eton College. During his long period in Italy as ambassador in Venice Wotton had not accumulated any great wealth and on returning to England he was moved to write a letter to The Duke of Buckingham in which he remarked: 'I am left utterly destitute of all possibility to subsist at home: Much like those seal-fishes, which sometimes, as they say, oversleeping themselves in ebbing water, feel nothing about them in a dry shore when they wake.'

Wotton's endeavours to secure the preferment had to be particularly strenuous since, as we have also previously noted, one of his rival contenders was the illustrious Sir Francis Bacon who was anxious to secure his own future having recently disgraced himself through accepting bribes and having suffered the ignominy of being imprisoned in the Tower and being required in addition to forfeit the astronomical sum of £40,000.

Wotton was aware that the Eton Provostship was essentially in the gift of George Villiers the first Duke of Buckingham, by repute amongst his contemporaries 'one of the handsomest men in the world, generous and brave' who was deeply interested in architecture and was a discerning collector of paintings and manuscripts. His likeness is known to us through the striking portrait of him by Anthony van Dyck. Wotton, the astute observer of the ways of men and the practised diplomat, decided to catch the attention of Buckingham by appealing to his aesthetic sensibilities. In all his time in Italy Wotton had been surrounded by works of art and architectural masterpieces of unrivalled splendour and profusion, the influence of which had worked deeply upon his intellectual and artistic inclinations. Accordingly, he calculated that a treatise upon architecture based upon the best authors and cast in his own adroit style of writing would make a favourable impression upon Buckingham and thereby strengthen his chances of being selected. And so was borne, through these somewhat unedifying machinations, the first theoretical treatise on architecture in the English language. It has become justly celebrated not least through its famous opening lines which are the most frequently quoted aphoristic passage to be found in any architectural writing.

Wotton produced his text at great speed having the pages printed as rapidly as he could compose them. The entire work was completed within two months of him having returned from Italy, a remarkable tribute to his powers of cultural assimilation, his discerning eye and memory for detail, his knowledge of authors on architecture, and his stylish dexterity in manipulating the English language.

Wotton's endeavours served him well since, as we have already remarked, he was successfully appointed to the Provostship and, more lastingly, generations have cause to be grateful for his endeavours in securing for posterity the rich gleanings of a well-informed mind.

The speed with which Wotton drafted his book precluded him from enlisting support for his enterprise in the form of a group of subscribers. He did nevertheless ensure that a circle of influential persons received his work which included King James I, the Prince of Wales, the Earl of Middlesex, the Archbishop of Canterbury, Sir William Boswell and his main target the Duke of Buckingham.

The Elements of Architecture is a small-scale work with a format of about 15cm x 23cm. It contains no illustrations or diagrams. The publication was undertaken by John Bill, the King's printer, who knew Wotton personally. In his Preface Wotton acknowledges his debt to others: 'I am but a gatherer and disposer of other mens stuffe, at my best value.' His primary authorities are Vitruvius, 'Our principal Master' and Leon-Batista Alberti '*the* Florentine, *whom I repute the first learned* Architect, *beyond the* Alpes.' In the course of his text, Wotton concedes his obligation to several other writers on and practitioners of architecture and to the work of philosophers and painters.

The book is in two parts. Part I is concerned with the art of Building: Wotton's view is direct and pragmatic: 'In *Architecture* as in all other *Operative* Arts, the *end* must direct the *Operation*.' and from this it follows 'The *end* is to build well'. Following this injunction, Wotton sets out the first part of his text under a series broad themes in which he describes the art of Building with particular reference to the site or situation, the principal parts, materials and architectural forms. In our presentation of Wotton's text, we have retained the orthography of the original including headings, sub-headings and the approximate size of letters and symbols. Wotton's layout has been adapted to more conveniently fit the format of our own work and additional subject-headings have been introduced which are designated by square brackets. By collecting these headings together we have in effect an inventory of Wotton's themes as listed below:

The Elements of Architecture: Part I

The principal themes in Wotton's text.

The *Seat*, and the *Worke*
The placing of the *Parts*
The *Materials* in the *Worke*
The *Figuration*
The outward *Aspect*
Concerning the *Parts*
The *Foundation*
The *Substruction*
The *Wals*
The *Intermissions*
The *Communities* of the *Orders*
The *Tuscan Order*
The *Dorique Order*
The *Ionique Order*
The *Corinthian Order*
The *Compound Order*
Cautions relating to the *Orders*
Pylasters
Arches
Theoremes
Apertions
Of *Doors* and *Windows*
Of *Staire-cases*
Of *Chimnies*
Conducts for the *Suillage*
Compartition: generall Precautions
The *Compartition*
Usefulnesse
The *Offices*
Lodging chambers
Floore or *Area*
Cover or *Roofe*

In Part II of his text Wotton concerns himself with the principles governing the adornment of buildings having regard first to the interiors of rooms and second to the use of decorative embellishments outside. He commences this text in the forthright manner that characterises the opening of Part I with a typical didactic flourish: 'Every Mans proper *Mansion* House and *Home*, being the *Theater* of his *Hospitality*, the *Seate* of *Selfe-fruition*, the *Comfortablest part* of his owne *Life*, the *Noblest* of his Sonnes *Inheritance*, a kinde of private *Princedome*; Nay, to the *Possessors* thereof, an *Epitomie* of the whole *World*: may well deserve by these *Attributes*, according to the degree of the *Master*, to be *decently* and *delightfully* adorned.' Part II is somewhat shorter than Part I but is similarly laid out in sections for which we have provided headings, a list of which follows:

The Elements of Architecture: Part II

The principal themes in Wotton's text.

Ornaments both *within* and *without* the *fabrique*
Picture and *Sculpture* in the *beautifying* of *Fabriques*
The disposing of *Pictures* within
Sculpture
The placing of *Sculpture*
Mosaique and *Plastique*
Garden *Ornaments*
Fountaines
Groves
Aviaries
Ponds
Methode for *Worke* to be *Commodious, Firme*, and *Delightfull*

Sir Henry Wotton's *The Elements of Architecture* deserves its high reputation and the affectionate regard in which it is widely held. Wotton was no mere mechanical editor-compiler and his text shimmers with intelligence and sparkles with witty, informed observation. His perspicacious style is clear, endearing and illuminating. In a relatively few pages he distils a large corpus of ideas that were nurtured through many years of direct observation of original source materials including buildings, artefacts and writings about them. His work inspired others to follow his example in the writing of architectural texts the primary purpose of which was to inform and to elevate taste through the inculcation of better and deeper understanding.

3 *The Elements of Architecture*

Title Page

The Preface

Wotton's Text:

Part I and Part II

THE ELEMENTS OF ARCHITECTVRE,

Collected by

HENRY WOTTON Knight, *from the best Authors* and Examples.

LONDON
Printed by IOHN BILL,
M.DC.XXIV.

THE

PREFACE.

I *Shall not neede (like the most part of Writers) to celebrate the* Subject *which I deliver. In that point I am at ease. For* Architecture, *can want no commendation, where there are* Noble Men, *or* Noble mindes; *I will therefore spend this* Preface, *rather about* those, *from whom I have gathered my knowledge; For I am but a gatherer and disposer of other mens stuffe, at my best value.*

Our principall Master *is* Vitruvius *and so I shall often call him; who had this* felicitie, *that he wrote when the* Roman Empire *was neere the pitch; Or at least, when* Augustus *(who favoured his endeavours) had some meaning (if he were not mistaken) to bound the* Monarchie: *This I say was his good happe; For in growing and enlarging times,* Artes *are commonly drowned in* Action: *But on the other side, it was in truth an* unhappinesse, *to expresse himselfe so ill, especially writing (as he did) in a season of the ablest* Pennes; *And* his obscuritie *had this stange fortune; That though he were best practised, and best followed by his owne* Countrymen; *yet after the reviving and repolishing of good* Literature, (*which the combustions and tumults of the* middle Age *had* uncivillized) *he was best, or at least, first understood by* strangers: *For of the* Italians *that tooke him in hand, Those that were* Gramarians *seeme to have wanted* Mathematicall *knowledge; and the* Mathematicians *perhaps wanted* Gramer: *till both were sufficiently conjoined, in* Leon-Batista Alberti *the* Florentine, *whom I repute the first learned* Architect, *beyond the* Alpes; *But hee studied more indeede to make himselfe an* Author, *then to illustrate his* Master. *Therefore among his* Commenters, *I must (for my private conceite) yeild the chiefe praise unto the* French, *in* Philander; *and to the high* Germans, *in* Gualterus Rivius: *who, besides his notes, hath likewise published the most elaborate* translation, *that I thinke is extant in any* vulgar *speech of the world: though not without bewayling, now and then, some defect of* Artificiall *tearmes in his* owne; *as I must likewise*; *For if the* Saxon, (*our* mother *tongue*) *did complaine*; *as justly* (*I doubt*) *in this point may the* Daughter: Languages, *for the most part in tearmes of* Art *and* Erudition, *retayning their originall povertie, and rather growing rich and abundant*, *in complementall phrases and such froth. Touching divers* moderne *men that have written out of meere practise*, *I shall give them their due*, *upon occassion*.

And now, *after this short* Censure *of others*, *I would faine satistfie an* Objection *or two*, *which seeme to lie somewhat heavily upon my selfe*; *It will be said that I handle an* Art, *no way suteable either to my* employments, *or to my* fortune. *And so I shall stand charged*, *both with* Intrusion, *and with* Impertinency.

To the First *I answere*, *that though by the ever acknowledged goodnesse of my most* deare *and* gracious SOVERAIGNE; *and by his long indulgent toleration of my defects. I have borne abroad some part of his* civill *service*; *yet when I came home*, *and was againe resolved into mine owne simplicitie*, *I found it fitter for my* Pen (*at least in this first publique adventure*) *to deale with these plaine* compilements, *and tractable* Materials; *then with the* Laberynthes *and* Mysteries *of* Courts *and* States; *And less presumption for mee, who have long contemplated a famous* Republique, *to write now of* Architecture; *then it was anciently for* Hippodamus *the* Milesian, *to write of* Republiques, *who was himselfe but an* Architect.

To the Second, *I must shrinke up my shoulders, as I have learn'd abroad, and confesse indeed, that my fortune is very unable to* exemplifie, *and* actuate *my* Speculations *in this* Art, *which yet in trueth, made mee the rather even from my very disabilitie, take encouragement to hope; that my present* Labour, *would find the more favour with others, since it was undertaken for no mans* sake, *lesse then mine* owne. *And with that confidence, I fel into these thoughts; Of which, there were two wayes to be delivered; The one* Historical, *by description of the principall* workes, *preformed already in good part, by* Giorgio Vassari *in the lives of* Architects: *The other* Logicall, *by casting the* rules *and cautions of this Art, into some comportable* Methode: *whereof I have made choice; not onely as the shortest and most* Elementall; *but indeed as the* soundest. *For though in practicall* knowledges, *every complete* example, *may beare the credite of a* rule; *yet peradventure* rules *should precead, that we may by them, be made fit to judge of* examples: *Therefore to the* purpose; *for I will preface no longer.*

O F
THE ELEMENTS
O F
ARCHITECTURE.

The I. part.

IN *Architecture* as in all other *Operative* Arts, the *end* must direct the *Operation*.

The *end* is to build well.

Well building hath three Conditions.

Commoditie, *Firmenes*, and *Delight*.

A common division among the Deliverers of this *Art*, though I know not how, some what misplaced by *Vitruvius* himselfe *lib*. 1. *cap*. 3. whom I shalbe willinger to follow, as a Master of *Proportion*, then of *Methode*.

Now, For the attayning of these *Intentions*, wee may consider the whole *Subject* under two generall Heads.

The *Seate*, and the *Worke*.

Therefore first touching *Scituation*.

The *Precepts* thereunto belonging, doe either concerne the *Totall Posture*, (as I may tearm it) or the *Placing* of the *Parts*: wherof the first sort, howsoever usually set downe by *Architects* as a piece of their *Profession*: yet are in truth borrowed, from other *Learnings*: there being betweene *Arts* and *Sciences*, as well as betweene *Men*, a kind of good fellowship, and communication of their *Principles*.

For you shall finde some of them, to be meerely *Physicall*, touching the quality and temper of the *Aire*: which being a perpetuall ambient, and ingredient, and the defects thereof, incorrigible in single *Habitations* (which I most intend) doth in those respects, require the more exquisite caution; That it be not too *grosse*, nor too *penetrative*; Not subject to any foggy noysomnesse, from *Fenns* or *Marshes* neere adjoining; nor too *Mineral* exhalations, from the Soile it selfe. Not undigested, for want or *Sunne*, Not unexercised, for want of *Winde*; which were to live (as it were) in a *Lake*, or standing *Poole* of *Aire*, as *Alberti*, the *Florentin Architect*, doth ingeniously compare it.

Some doe rather seeme a little *Astrological*, as when they warne us from Places of malign *Influence*: where *Earth-quakes*, *Contagions*, *Prodigious Births*, or the like, are frequent without any evident cause: whereof the Consideration is peradventure not altogether vaine: Some are plainely *Oecomonical*; As that the *Seate* be well watered, and well fewelled, That it bee not of too steepie and incommodious *Accesse* to the trouble both of friends and *familie*. That it lie not too farre, from some navigable *River* or *Arme* of the *Sea*, for more ease of provision and such other *Domestique* notes.

Some againe may bee said to bee *Optical*? Such I meane as concerne the *Properties* of a well chosen *Prospect*: which I will call the *Royaltie* of *Sight*. For as there is a *Lordship* (as it were) of the *Feete*, wherein the Master doth much joy when he walketh about the *Line* of his owne *Possessions*: So there is a *Lordship* likewise of the *Eye* which being a raunging, and Imperious, and (I might say) an *usurping Sence*; can indure no narrow *circumscription*; but must be fedde, both with extent and varietie. Yet on the other side, I finde vaste and indefinite views which drowne all apprehension of the uttermost *Objects*, condemned, by good Authors, as if thereby some part of the pleasure (whereof I speake) did perish. Lastly, I remember a private *Caution*, which I know not well how to sort, unlesse I should call it *Political*. By no meanes, to build too neere a great *Neighbour*; which were in truth to bee as unfortunately seated on the earth, as *Mercurie* is in the Heavens, for the most part, ever in *combustion*, or *obscuritie*, under brighter beames then his owne.

From these several *Knowledges* as I have said, and perhaps from some other doe *Architects* derive their Doctrine about electon of *Seats*: wherein I have not beene so severe, as a great Scholer of our time, who precisely restrayneth a perfect *Scituation*, at least for the maine point of health, *Ad locum contra quem Solradios suos fundit cum sub Ariete oritur*, That is, in a word hee would have the first *salutation* of the *Spring*. But such *Notes* as these, wheresoever we finde them in grave or slight Authors, are to my conceite rather *wishes* then *Precepts*; and in that qualitie, I will passe them over. Yet I must withall say that in the *seating* of our selves (which is a kinde of *Marriage* to a *Place*) *Builders* should bee as circumspect as *Wooers*; lest when all is done that *Doome* befall us, which our Master doth lay upon *Mitylene*: *A Towne* in truth (saith hee) *finely built, but foolishly planted.* And so much touching that, which I termed the *Totall Posture*.

[The placing of the *Parts*.]

The next in Order is the placing of the *Parts*; About which (to leave as little as I may in my present labour, unto *Fancie*, which is wilde and irregular) I will propound a Rule of mine owne Collection, upon which I fell in this maner. I had noted, that all *Arte* was then in truest perfection, when it might bee reduced to some naturall *Principle*. For what are the most judicious *Artisans* but the *Mimiques* of *Nature*? This led me to contemplate the Fabrique of our owne Bodies, wherein the *High Architect* of the world, had displaied such skill, as did stupifie, all humane reason. There I found the *Hart* as the fountaine of Life placed about the Middle, for the more equall communication of the vitall spirits. The *Eyes* seated aloft, that they might describe the greater Circle within their view. The *Armes* projected on each side, for ease of reaching. Briefly (not to loose our selves in this sweet speculation) it plainely appeareth, as a Maxime drawne from the Divine light; That the *Place* of every part, is to be determined by the *Use*.

So then, from naturall *structure*, to proceed to Artificiall; and in the rudest things, to preserve some *Image* of the excellentest. Let all the principall chambers of *Delight*, All *Studies* and *Libraries*, be towards the *East*: For the Morning is a friend to the Muses. All Offices that require heat, as *Kitchins*, *Stillatories*, *Stoves*, roomes for *Baking*, *Brewing*, *Washing*, or the like, would be *Meridionall*. All that need a coole and fresh temper, as *Cellers*, *Pantries*, *Butteries*, *Granaries*, to the *North*. To the same side likewise, all that are appointed for gentle Motion, as *Galleries*, especially in warme Climes, or that otherwise require a steadie and unvariable light, as *Pinacothecia* (saith *Vitruvius*) by which he intendeth, (if I may guesse at his Greeke, as wee must doe often even at his Latine) certain *Repositories* for workes of rarity in Picture or other Arts, by the Italians called *Studioli*, which at any other Quarter, where the course of the *Sunne* doth diversifie the *Shadowes*, would loose much of their grace. And by this Rule having alwayes regarde to the *Use*, any other Part may bee fitly accommodated.

I must here not omit to note that the Ancient *Grecians*, and the *Romanes* by their example in their buildings abroad, where the *Seat* was free, did almost Religiously scituate the *Front* of their houses, towards the South; perhaps that the Masters *Eye*, when hee came home, might not be dazeled, or that being illustrated, by the *Sunne*, it might yeeld the more gracefull *Aspect*; or some such reason. But from this, the Moderne Italians doe varie, whereof I shall speake more in another place. Let thus much suffice at the present for the *Position* of the several *Members*, wherein must bee had as our Author doth often insinuate, and especially *lib*. 6. *cap*. 10. a singular regard, to the nature of the *Region*: Every Nation, being tyed above al Rules whatsoever, to a discretion, of providing against their owne *Inconveniences*: And therefore a good *Parler* in Ægypt would perchance make a good *Celler* in England.

There now followeth the second Branch of the generall *Section* touching the *Worke*.

[The *Materials* in the *Worke*.]

In the *Worke*, I will first consider the principall parts, and afterwards the Accessorie, or *Ornaments*; And in the Principall, first the Preparation of the *Materials*, and then the Disposition, which is the *Forme*.

Now, concerning the *Materiall* part; Although surely, it cannot disgrace an *Architect*, which doth so well become a Philosopher, to looke into the properties of *Stone* and *Wood*: as that Firre Trees, Cypresses, Cedars, and such other *Aereall* aspiring *Plants*, being by a kinde of naturall rigour (which in a man I would call pride) inflexible downewards are thereby fittest for *Posts* or *Pillars* or such upright use; that on the other side, Oake, and the like true hartie Timber being strong in all positions, may bee better trusted in crosse and traverse worke, for *Summers*, or guirding and binding *beames*, as they tearme them. And so likewise to observe of *Stone*, that some, are better within, and other to beare *Weather*: Nay, to descend lower even to examine *Sand* and *Lyme*, and *Clay* (of all which things *Vitruvius* hath discoursed, without any daintines, and the most of new Writers) I say though the *Speculative* part of such knowledge be *liberall*. Yet to redeeme this profession, and my present paynes, from indignities, I must heere remember that to choose and sort the *materials*, for every part of the *Fabrique*, is a Dutie more proper to a second *Superintendent*, over all the Under Artisans called (as I take it) by our Author, *Officinator lib*. 6. *cap*. 11. and in that Place expressely distinguished, from the *Architect*, whose glory doth more consist, in the Designement and *Idea* of the whole *Worke*, and his truest ambition should be to make the *Forme*, which is the nobler Part (as it were) triumph over the *Matter*: whereof I cannot but mention by the way, a foreigne Paterne, namely the Church of *Santa Giustina* in *Padoua*: In truth a sound piece of good Art, where the *Materials* being but ordinarie stone, without any garnishment of sculpture, doe yet ravish the Beholder, (and hee knowes not now) by a secret *Harmony* in the *Proportions*. And this indeede is that end, at which in some degree, we should ayme even in the privatest workes: whereunto though I make haste, yet let me first collect, a few of the least trivial cautions, belonging to the *Materiall Provision*.

Leon Batista Alberti, is so curious, as to wish all the *Timber*, cut out of the same *Forrest*, and al the *Stone*, out of the same *Quarrie*.

Philibert de l'Orme the French Architect goes yet somewhat further, and would have the *Lyme* made of the very same *Stone*, which wee intend to employ in the *Worke*; as belike imagining that they will sympathize and joyne the better, by a kinde of *Originall* kindred. But such conceits as these seeme somewhat too fine among this *Rubbage*, though I doe not produce them in sport. For surely the like agreements of nature, may have oftentimes a discreet application to Art. Alwayes it must be confessed, that to make *Lyme* without any great choyce of refuse stuffe, as we commonly do, is an *English* error, of no small moment in our Buildings. Whereas the *Italians* at this day, and much more the *Ancients* did burn their firmest *stone*, and even fragments of *Marble* where it was copious, which in time became almost *Marble* againe, or at least of indissoluble duritie, as appeareth in the standing *Theaters*. I must here not omit, while I am speaking of this part, a certain forme of *Bricke* described by *Daniele Barbaro Patriarch* of *Aquileia*, in the largest Edition of his Commentary upon

Vitruvius. The Figure triangular, every side a foot long, and some inch and a halfe thicke, which he doth commend unto us for many good conditions: As that they are more commodious in the management, of lesse expence, of fayrer show, adding much beautie and strength to the *Murall Angles*, where they fall gracefully into an indented *Worke*: so as I should wonder that wee have not taken them into use, being propounded by a Man of good authoritie in this knowledge; but that all Nations doe start at *Novelties*, and are indeed maried to their owne *Moulds*. Into this place might aptly fall a doubt, which some have wel moved; whether the ancient *Italians* did burne their *Bricke* or no; which a passage or two in *Vitruvius* hath left ambiguous. Surely where the *Naturall* heat is strong enough, to supply the *Artificiall*, it were but a curious folly to multiply both *Labour* and *Expence*. And it is besides very probable, that those *Materials* with a kindely and temperate heate would prove fairer, smoother, and lesse distorted, then with a violent: Onely, they suffer two exceptions. First, that by such a gentle drying much time will bee lost which might otherwise bee employed in compiling. Next, That they will want a certaine sucking and soaking *Thirstinesse*, or a fiery appetite to drinke in the *Lime*, which must knit the *Fabrique*. But this question may be confined to the *South*, where there is more Sunne and patience. I will therefore not hinder my course, with this incident scruple, but close that part which I have now in hand, about the *Materialls*, with a principall caution: That sufficient *Stuffe* and *Money* bee ever ready before we beginne: For when wee build now a piece, and then another by *Fits*, the *Worke* dries and sinkes unequally, whereby the *Walles* growe full of *Chinques*, and *Crevices*; Therefore such pawsings are well repooved by *Palladio*, *lib*. 1. *cap*. 1. and by all other. And so having gleaned these few remembrances, touching the preparation of the *Matter*, I may now proceede to the *Disposition* thereof, which must forme the *Worke*. In the

Forme, as I did in the *Seate*, I will first consider the generall *Figuration*, and then the severall *Members*.

[The *Figuration*.]

Figures are either *simple* or *Mixed*. The simple be either *Circular* or *Angular*. And of Circular, either *Compleate*, or *Deficient*, as *Ovals*, with which kindes I will bee contented, though the Distribution might bee more curious.

Now the exact *Circle* is in truth a Figure, which for our purpose hath many fit and eminent properties; as fitnesse, for Commodity and Receit, being the most capable; fitnesse for strength and duration, being the most united in his parts; Fitnesse for beautie and delight, as imitating the celestiall *Orbes*, and the universall *Forme*. And it seemes, besides, to have the approbation of *Nature*, when shee worketh by *Instinct*, which is her secret Schoole: For birds doe build their nests *Spherically*: But notwithstanding these Attributes, it is in truth a very unprofitable Figure in private *Fabriques*, as being of all other the most chargeable, and much roome lost in the bending of the *Walles*, when it comes to bee divided: besides an ill distribution of light, except from the *Center* of the *Roofe*. So as anciently it was not usual, save in their *Temples* and *Amphi-Theaters*, which needed no Compartitions. The *Ovals* and other imperfect circular Formes, have the same exceptions, and lesse benefite of capacity: So as there remaynes to bee considered in this generall survey of *Figures*, the *Angular*, and the *Mixed* of both. Touching the *Angular*, it may perchance sound somewhat strangely, but it is a true observation, that this *Art* doth neither love many *Angles* nor few. For first, the *Triangle* which hath the fewest sides and corners is of all other the most condemned, as being indeed both incapable and infirme (wherof the reason shall be afterwards rendred) and likewise unresolvable into any other regular *Forme* then it selfe, in the inward *Partitions*.

As for Figures of five, six, seven, or more *Angles*; They are surely fitter for *Military Architecture*, where the Bulworks may be layed out at the *Corners*, and the sides serve for *Curtaines*, then for civill use; though I am not ignorant of that famous Piece at *Caprarola*, belonging to the house of *Farnese*, cast by *Baroccio* into the forme of a *Pentagone*, with a *Circle* inscribed, where the *Architect* did ingeniously wrestle with divers inconveniences in disposing of the *Lights*, and in saving the vacuities. But as designes of such nature doe more ayme at Rarity, then Commoditie: so for my part I had rather admire them, then commend them.

These things considered, we are both by the Precepts and by the Practise of the best Builders, to resolve upon *Rectangular Squares*, as a meane betweene too few, and too many Angles; and through the equall inclination of the sides (which make the right Angle) stronger then the *Rhombe*, or *Losenge*, or any other irregular *Square*. But whether the exact *Quadrat*, or the long *Square* be the better, I finde not well determined, though in mine owne conceit I must preferre the latter; provided that the *Length* doe not exceed the *Latitude* above one third part, which would diminish the beauty of the *Aspect*, as shall appeare when I come to speake of *Symmetry* and *Proportion*.

Of mixed Figures, partly *Circular*, and partly *Angular*, I shall neede to say nothing; because having handled the simple already, the mixed according to their composition, doe participate of the same respects. Onely against these, there is a proper *Objection*, that they offend *Uniformity*: Whereof I am therefore opportunely induced to say somewhat, as farre as shal concerne the outward *Aspect*, which is now in Discourse.

[The outward *Aspect*]

In *Architecture*, there may seem to be two opposite affectations, *Uniformitie* and *Varietie*, which yet will very well suffer a good reconcilement, as we may see in the great Paterne of *Nature*, to which I must often resort: For surely there can be no *Structure*, more uniforme, then our *Bodies* in the whole *Figuration*: Each side, agreeing with the other, both in the number, in the qualitie, and in the measure of the Parts: And yet some are round, as the *Armes*, some flat, as the *Hands*, some prominent, and some more retired: So as upon the Mater, wee see that *Diversitie* doth not destroy *Uniformitie*, and that the Limmes of a noble *Fabrique*, may bee correspondent enough, though they be various; Provided alwayes, that we doe not runne into certaine extravagant Inventions, whereof I shall speake more largely, when I come to the parting and casting of the whole *Worke*. We ought likewise to avoyde Enormous heights of sixe of seven *Stories*, as well as irregular *Formes*; and the contrary fault of low-distended *Fronts*, is as unseemely: Or againe, when the *Face* of the Building, is narrow and the *Flank* deepe; To all which extreames, some particular Nations, or Townes, are subject, whose Names may be civilly spared: And so much for the generall *Figuration*, or *Aspect of* the *Worke*.

[Concerning the *Parts*.]

Now concerning the Parts in Severalitie. All the parts of every *Fabrique*, may be comprised under five Heads, which Division I receive from *Batista Alberti*, to doe him right. And they be these.

The *Foundation*.
The *Walles*.
The *Appertions or Overtures*.
The *Compartition*.
And the *Cover*.

About all which I purpose to gather the principall Cautions, and as I passe along, I will touch also the naturall Reasons of *Art*, that my discourse may be the lesse *Mechanicall*.

[The *Foundation.*]

First then concerning the *Foundation*, which requireth the exactest care; For if that happen to dance, it will marre all the mirth in the House: Therefore, that wee may found our Habitation firmely, we must first examine the *Bed* of *Earth* (as I may tearme it,) upon which we will Build; and then the underfillings, or *Substruction*, as the Auncients did call it: For the former, we have a generall precept in *Vitruvius* twice precisely repeated by him, as a point indeed of mayne consequence, first *lib.* 1. *cap.* 5. And againe more fitly *lib.* 3. *cap.* 3. in these words as *Philander* doth well correct the vulgar Copies.

Substructionis Fundationes fodiantur (saith he) *si queant inueniri ad solidum, etc. in solido.* By which words, I conceive him to commend unto us, not onely, a diligent, but even a jealous examination what the *Soile* will beare: advising us, not to rest upon any appearing *Soliditie*, unlesse the whole *Mould* through which wee cut, have likewise beene *solid*; But how deepe wee should goe in this search, hee hath no where to my remembrance determined, as perhaps depending more upon *Discretion*, then *Regularitie*, according to the weight of the *Worke*; yet *Andrea Palladio* hath fairely adventured to reduce it into Rule: Allowing for that *Cauazione* (as he calleth it) [Underdiging or, Hollowing of the Earth] a fixt part of the *height* of the whole *Fabrique*, unless the Cellers be under ground, in which case hee would have us, (as it should seeme) to found somewhat lower.

Some *Italians* doe prescribe, that when they have chosen the *Floore*, or Plot, and laid out the Limits of the *Worke*, wee should first of all Digge *Wels* and *Cesternes*, and other under-counducts and conveiances, for the *Suillage* of the House, whence may arise a double benefit, for both the Nature of the *Mould* or *Soile*, would thereby be safely searched, and moreover those open vents, will serve to discharge such *Vapours*, as having otherwise no issue might peradventure shake the

Building. This is enough for the naturall *Grounding*, which though it bee not a part of the solid *Fabrique*, yet here was the fittest place to handle it.

[The *Substruction.*]

There followeth the *Substruction*, or Ground-worke of the whole *Edifice*, which must sustaine the *Walles*; and this is a kinde of *Artificiall* foundation, as the other was *Natural*. About which these are the chiefe Remembrances. First, that the bottome be precisely levell, where the *Italians* therefore commonly lay a platforme of good Bord; Then that the lowest *Ledge* or *Row* be meerely of *Stone*, and the broader the better, closely layd without *Morter*, which is a generall caution for all parts in Building, that are contiguous to *Bord* or *Timber*, because *Lime* and *Wood* are insociable, and if any where unfit confiners, then most especially in the *Foundation*. Thirdly, that the bredth of the *Substruction* bee at least double to the *insistent Wall*, and more or lesse, as the weight of the *Fabrique* shall require; for as I must againe repeate, *Discretion* may be freer then *Art*. Lastly, I finde in some a curious precept, that the *Materials* below, be layd as they grew in the *Quarrie*, supposing them belike to have most strength in their *Naturall* and *Habituall* Posture. For as *Philippe de l'Orme* observeth, the breaking or yeelding of a stone in this part, but the bredth of the backe of a knife, will make a *Cleft* of more then half a foot in the *Fabrique* aloft, So important are *Fundamentall errors*. Among which notes I have sayd nothing of *Pallification*, or *Pyling* of the *Ground-plot*, commanded by *Vitruvius*, when we build upon a moist or marshy soile, because that were an error in the first choyce. And therefore all *Seats* that must use such provision below (as *Venice* for an eminent example) would perhaps upon good enquiry, be found to have beene at first chosen by the counsell of *Necessity*.

Now the *Foundation* being searched, and the *Substruction* lay'd, wee must next speake of the *Wals*.

[The *Wals*.]

Wals are either entire and continuall, or intermitted; and the *Intermissions* be either *Pillars* or *Pylasters*, for here I had rather handle them, then as some others doe, among *Ornaments*.

The entire *Muring* is by Writers diversly distinguished: By some, according to the quality of the *Materials*, as either Stone or Brick, *etc*. where, by the way, let me note, that to build *Wals* and greater Workes of *Flint*, whereof wee want not example in our *Iland*, and particularly in the *Province* of *Kent*, was (as I conceive) meerely unknown to the *Ancients*, who observing in that *Materiall*, a kinde or *Metalicall* Nature, or at least a *Fusibility*, seeme to have resolved it into nobler use; an Art now utterly lost, or perchance kept up by a few *Chymicks*. Some againe doe not so much consider the quality, as the *Position* of the sayd *Materials*: As when Bricke or squared stones are laid in their lengths with sides and heads together, or their points conjoyned like a *Networke* (for so *Vitruvius* doth call it *reticulatum opus*) of familiar use (as it should seeme) in his Age, though afterwards growne out of request, even perhaps for that subtill speculation which hee himselfe toucheth; because so layd, they are more apt in swagging down, to pierce with their points, then in the jacent Posture, and so to crevice the *Wall*: But to leave such cares to the meaner Artificers, the more essentiall are these.

That the *Walles* bee most exactly perpendicular to the *Ground-worke*: for the right *Angle* (thereon depending) is the true cause of all *Stability*; both in Artificiall and Naturall positions; A man likewise standing firmest, when he stands uprightest. That the massiest and heaviest *Materials* bee the lowest, as fitter to beare, then to be borne. That the *Worke* as it riseth, diminish in *thicknesse* proportionally, for ease both of weight, and of expence. That certaine courses or *Ledges* of more strength then the rest, be interlayed like *Bones*, to sustaine the *Fabrique* from totall ruine, if the under parts should decay. Lastly, that the *Angles* bee firmely bound, which are the *Nerves* of the whole *Edifice*, and therefore are commonly fortified by the *Italians*, even in their Bricke buildings, on each side of the corners, with well squared stone, yeelding both strength and grace And so much touching the entire or solid *Wall*.

[The *Intermissions*.]

The intermissions (as hath beene sayd) are either by *Pillars*, or *Pylasters*.

Pillers which we may likewise call *Columnes* (for the word among Artificers is almost naturallized) I could distinguish into a *Simple* and *Compounded*. But (to tread the beaten and plainest way) there are five *Orders* of Pillers, according to their dignity and perfection, thus marshalled:

The *Tuscan*.
The *Dorique*.
The *Ionique*.
The *Corinthian*.

And the *Compound Order*, or as some call it the *Roman*, others more generally the *Italian*.

In which five *Orders* I will first consider their *Communities*, and then their *Proprieties*.

[The *Communities* of the *Orders*.]

Their *Communities* (as farre as I observe) are principally three. First, they are all *Round*; for though some conceive *Columna Atticurges* mentioned by *Vitruvius*, *lib*. 3. *cap*. 3. to have been a squared Pillar, yet wee must passe it over as irregular, never received among these *Orders*, no more then certaine other licentious inventions, of *Wreathed*, and *Vined*, and *Figured Columnes*, which our Author himselfe condemneth, being in his whole Booke a proffesed enemy to *Fancies*.

Secondly, they are all *Diminished*, or *Contracted* insensibly, more or lesse, according to the proportion of their heights, from one third part of the whole *Shaft* upwards, which *Philander* doth prescibe by his owne precise measuring of the Ancient remainders, as the most gracefull *Diminution*. And here I must take leave to blame a pratice growne (I know not how) in certaine places too familiar, of making *Pillars* swell in the middle, as if they were sicke of some *Tympany*, or *Dropsie*, without any Authentique Paterne or Rule, to my knowledge, and unseemly to the very judgement of sight. True it is that in *Vitruvius*, *lib*. 3. *cap*. 2. wee finde these words, *De adjectione*, *quæ adijcitur in medijs Columnis*, *quæ apud Grecos appellatur*, *in extremo libro erit formatio eius*; which passage, seemeth to have given some countenance to this error. But of the promise there made, as of diverse other elsewhere, our Master hath fayled us, either by slip of memory, or injury of time, and so wee are left in the darke. Alwayes sure I am, that besides the authority of example which it wanteth, It is likewise contrary to the Originall and Naturall *Type*, in *Trees*, which at first was imitated in *Pillars*, as *Vitruvius* himselfe observeth, *lib*. 5. *cap*. 1. For who ever saw any *Cypresse*, or *Pine*, (which are there alledged) small below and above, and tumerous in the middle, unlesse it were some diseased Plant, as Nature (though otherwise the comliest *Mistresse*) hath now and then her deformities and Irregularities.

Thirdly, they have all their *Under-settings*, or *Pedistals*, in height a third part of the whole *Columne*, comprehending the *Base* and *Capitall*; and their upper Adjuncts, as *Architrave*, *Frize*, and *Cornice*, a fourth part of the sayd Pillar; which rule of singular use and facility I find setled by *Jacobo Baroccio*, and hold him a more credible Author, as a man that most intended this piece, then any that vary from him in those *Dimentions*.

These are their most considerable *Communities* and agreements.

Their *Proprieties* or *Distinctions* will best appeare by some reasonable description of them all, together with their *Architraves*, *Frizes*, and *Cornices*, as they are usually handled.

[The *Tuscan Order*.]

First therefore the *Tuscan* is a plain, massie, rurall Pillar, resembling some sturdy well-limmed Labourer, homely clad, in which kinde of comparison *Vitruvius* himselfe seemeth to take pleasure, *lib*. 4. *cap*. 1. The length thereof shall be six *Diameters*, of the grossest of the Pillar below. Of all proportions, in truth, the most naturall; For our Author tells us, *lib*. 3. *cap*. 1. that the foote of a man is the fixt part of his bodie in ordinary measure, and *Man* himselfe, according to the saying of *Protagoras* (which *Aristotle* doth somewhere vouchsafe to celebrate) is as it were the *Prototype* of all exact *Symmetrie*, which we have all other occasion to touch before: This *Columne* I have by good warrant called *Rurall*, *Vitru*. *cap*. 2. *lib*. 3. And therefore we need not consider his rank among the rest. The distance or *Intercolumniation* (which word Artificers doe usually borrow) may bee neere foure of his owne *Diameters*, because the *Materials* commonly layd over this Pillar, were rather of wood then stone; through the lightnesse whereof the *Architrave* could not suffer, though thinnely supported, nor the *Columne* it selfe being so substantiall. The *Contraction* aloft shall be (according to the most received practice) one fourth part of his thicknesse below. To conclude, (for I intend only as much as shall serve for a due Distinguishment, and not to delineate every petty member) the *Tuscan* is of all the rudest Pillar, and his principall Character *Simplicity*.

[The *Dorique Order.*]

The *Dorique Order* is the gravest that hath beene received into civill use, preserving, in comparison of those that follow, a more *Masculine Aspect*, and little trimmer then the *Tuscan* that went before, save a sober garnishment now and them of *Lion's heads* in the *Cornice*, and of *Triglyphs* and *Metopes* alwayes in the *Frize*. Sometimes likewise, but rarely, chaneled, and a little slight sculpture about the *Hypotranchelion*, or *Necke* under the *Capitall*. The length, seven *Diameters*. His ranke or degree, is the lowest by all *Congruity*, as being more massie then the other three, and consequently abler to support. The *Intercolumniation*, thrice as much as his thicknesse below. The *Contraction* aloft, one fift of the same measure. To discerne him, will bee a peece rather of good *Heraldry*, then of *Architecture*: For he is best knowne by his place, when he is in company, and by the peculiar ornament of his *Frize* (before mentioned) when he is alone.

[The *Ionique Order.*]

The *Ionique Order* doth represent a kinde of Feminine slendernesse, yet saith *Vitruvius*, not like a light Housewife, but in a decent dressing, hath much of the *Matrone*. The length eight *Diameters*. In degree as in substantialnesse, next above the *Dorique*, sustayning the third, and adorning the second Story. The *Intercolumniation* two of his owne *Diameters*. The *Contraction* one sixt part. Best knowne by his trimmings, for the bodie of this *Columne* is perpetually chaneled, like a thicke plighted Gowne. The *Capitall* dressed on each side, not much unlike womens Wires, in a spirall wreathing, which they call the *Ionian Voluta*. The *Cornice* indented. The *Frize* swelling like a pillow; And therefore by *Vitruvius*, not unelegantly tearmed *Pulvinata*. These are his best *Characters*.

[The *Corinthian Order.*]

The *Corinthian*, is a *Columne*, laciviously decked like a Curtezane, and therein much participating (as all Inventions doe) of the place where they were first borne: *Corinthe* having been without controversie one of the wantonest Townes in the world. This *Order* is of nine *Diameters*. His degree, one Stage above the *Ionique*, and alwaies the highest of the simple *Orders*. The *Intercolumniation* two of his *Diameters*, and a fourth part more, which is of all other the comeliest distance. The *Contraction* one seventh Part. In the *Cornice* both *Dentelli* and *Modiglioni*. The *Frize*, adorned with all kinds of *Figures* and various Compartments at Pleasure [*Our Artizans call them Teeth and Cartonzes.*]. The *Capitall*, cut into the beautifullest leafe, that Nature doth yeeld, which surely next the *Aconitum Pardalianches* (rejected perchance as an ominous *Plant*) is the *Acanthus or Branca Ursina* though *Vitruvius* doe impute the choice thereof unto Chance, and wee must be contented to beleeve him: In short, As *Plainenesse* did Charactarize the *Tuscan*, so must *Delicacie* and *Varietie* the *Corinthian* Pillar, besides the height of his Ranke.

[The *Compounded Order.*]

The last is the *Compounded Order*: His *name* being a briefe of his *Nature*. For this Pillar is nothing in effect, but a *Medlie*, or an *Amasse* of all the precedent *Ornaments*, making a new kinde, by stealth, and though the most richly tricked, yet the poorest in this, that he is a borrower of all his Beautie. His length, (that he may have somewhat of his owne) shalbe of ten *Diameters*. His degree should, no doubt, bee the highest by reasons before yeelded. But few *Palaces* Auncient or Moderne exceede the third of the Civill *Orders*. The *Intercolumniation*, but a *Diameter* and an halfe, or alwayes somewhat less then two. The *Contraction* of this Pillar must be one eight Part lesse above then belowe. To know him will be easie by the verie mixture of his *Ornaments*, and *Cloathing*.

And so much touching the five *Orders* of *Columnes*, which I will conclude with two or three, not impertiment Cautions:

[*Cautions* relating to the *Orders*.]

First, that where more of these *Orders* then one, shalbe set in severall *Stories* or *Contignations*, there must bee an exquisite care, to place the *Columnes* precisely, one over another, that so, the *solid* may answere to the *solid*, and the *vacuities* to the *vacuities*, as well for *Beautie*, as strength of the *Fabrique*: And by this Caution the Consequence is plaine, that when wee speak of the *Intercolumniation* or *distance*, which is due to each *Order*, we meane in a *Dorique*, *Ionicall*, *Corinthian Porch*, or *Cloister*, or the like of one *Contignation*, and not in Storied buildings.

Secondly, let the *Columnes* above be a fourth part lesse then those below, saith *Vitruvius*, *lib*. 5. *cap*. 1. A strange Precept, in my opinion, and so strange, that peradventure it were more sutable, even to this owne Principles, to make them rather a fourth Part greater, For *lib*. 3. *cap*. 2. where our master handleth the *Contractions* of Pillars, wee have an *Optique* Rule, that the higher they are, the lesse should be alwayes their diminution aloft, because the *Eye* it selfe, doth naturally contract all *Objects* more or lesse, according to the *Distance*; which consideraton, may, at first sight, seeme to have beene forgotten in the Caution wee have now given; but *Vitruvius* (the best Interpreter of himselfe) hath in the same place of his fift Booke, well acquitted his memorie by these words: *Columnæ superiores quarta parte minores*, *quam inferiores*, *sant constituendæ*; *proptered quod*, *operi ferendo quæsunt inferiora*, *firmiora esse debent*; prefferring like a wise Mechanick, the *naturall* Reason, before the *Mathematicall*, and sensible conceits before abstracted. And yet *lib*. 4. *cap*.

4. he seemeth againe, to affect *Subtiltie*, allowing pillars the more they are chaneled, to bee the more slender; because while our *Eye* (saith hee) doth as it were distinctly measure, the eminent and the hollowed Parts, the Totall Object appeareth the bigger, and so as much as those excavations, doe subtract, is supplied by a *Fallacie* of the *Sight*: But here mee thinks, our Master should likewise have rather considered, that naturall Inconvenience; for though Pillars by chaneling, bee seemingly ingrossed to our Sight, yet they are truely weakened in themselves; and therefore ought perchance in sound reason not to bee the more slender, but the more Corpulent, unlesse apparances preponder truths, but *Contra Magistrum, non est disputandum.*

A third Caution shalbe that all the *projected* or *Jutting* Parts (as they are tearmed) be very moderate, especially, the *Cornices* of the lower *Orders*, for whilest some thinke to give them, a beautifull and royall Aspect, by their largenesse, they sometimes hinder both the Light within, (whereof I shall speake more in due place) and likewise detract much from the viewe of the Front without, as well appeareth in one of the principall Fabriques at *Venice*, namely the Palace, of the Duke *Grimani* of the *Canal Grande*, which by this magnificent errour, is somewhat *disgraced*: I neede now say no more concerning *Columnes* and their *Adjuncts*, about which Architects make such a noyse in their Bookes, as if the very tearmes of *Architraves*, and *Frizes*, and *Cornices*, and the like, were enough to graduate a Master of this *Art*; yet let me before I passe to other matter, prevent a familiar Objection; It will perchance bee said, that all this Doctrine touching the five *Orders*, were fitter for the *Quarries* of *Asia* which yeelded 127 *Columnes* of 60 *Foote* high, to the *Ephesian* Temple, or for *Numidia* where *Marbles* abound; then for the *Spirits* of *England*, who must be contented with more ignoble *Materials*: To which I answere, that this neede not discourage us: For I have often at *Venice* viewed with much pleasure, an

Atrium Græcum (we may translate it an *Anti porch*, after the Greeke manner) raised by *Andrea Palladio*, upon eight *Columnes* of the *Compounded Order*; the *Bases* of Stone, without *Pedistals*, the *shafts* or Bodies, of meere Brick; three foote and an halfe thick in the *Diameter* below, and consequently thirty five foote high, as himselfe hath described them in his second Booke; Then which, mine Eye, hath never yet beheld any *Columnes*, more stately of Stone or Marble; For the Bricks, having first beene formed in a *Circular Mould*, and then cut before their burning into foure quarters or more, the sides afterwards joyne so closely, and the points concenter so exactly, that the *Pillars* appeare one *entire Peece*; which short description, I could not omit, that thereby may appeare, how in truth wee want rather *Art* then stuffe, to satisfie our greatest *Fancies*.

[*Pylasters.*]

After *Pillars*, the next in my distribution, are *Pylasters*, mentioned by *Vitruvius*, *lib*. 5. *cap*. 1. and scant any where else under the name of *Parastates*, as *Philander* conceiveth, which *Grammaticall* point (though perchance nor very cleere) I am contented to examine no farther. Alwayes, what we meane by the thing it selfe, is plaine enough in our owne vulgar; Touching which, I will briefly collect the most considerable notes.

Pylasters, must not bee too tall and slender, least they resemble *Pillars*, nor too *Dwarfish* and grosse, least they imitate the *Piles* or *Peers* of Bridges: Smoothnesse doth not so naturally become them, as a Rustick *Superfices*, for they ayme more at *State* and *Strength*, then *Elegancie*. In private Buildings they ought not to be narrower, then one third, nor broader then two parts of the whole Vacuity, between *Pylaster* and *Pylaster*; but to those that stand at the *Corners*, may be allowed a little more Latitude by discretion, for strength of the *Angles*: In *Theaters* and *Amphi-theaters*, and such weighty

Workes, *Palladio* observeth them to have beene as broad as the halfe, and now and then as the whole *Vacuitie*: Hee noteth likewise (and others consent with him) that their true *Proportion*, should bee an *exact Square*; But for lessening of expence, and inlarging of roome, they are commonly narrower in *Flanke*, then in *Front*: Their principall *Grace* doth consist in halfe or whole *Pillars*, applied unto them; in which case it is well noted by Authors, that the *Columnes* may bee allowed somewhat above their ordinary length, because they leane unto so good *Supporters*. And thus much shall sufice touching *Pilasters*, which is a cheape, and a strong, and a noble kinde of *Structure*.

[*Arches.*]

Now because they are oftner, both for Beauty and Majesty, found *Arched*, then otherwise; I am heere orderly led to speake of *Arches*, and under the same head of *Vaults*: for an *Arch* is nothing indeed but a contracted *Vault*, and a *Vault* is but a dilated *Arch*: Therefore to handle this Piece both compendiously, and fundamentally, I will resolve the whole businesse in to a few *Theoremes*.

Theoreme 1.

All solid *Materials* free from impediment, doe descend *perpendicularly* downewards, because *ponderosity* is a naturall inclination to the *Center* of the World, and *Nature* performeth her motions by the shortest lines.

Theoreme 2.

Brickes moulded in their ordinary *Rectangular* Forme, if they shall be layd one by another in a levell row, betweene any Supporters sustayning the two ends; then all the pieces between, will necessarily sinke, even by their owne naturall *Gravity*, and much more if they suffer any depression by other waight above them, because their sides being *parallel*, they have roome to descend *perpendicularly*, without impeachment, according to the former *Theoreme*; Therefore to make them stand, wee must either change their *Posture*, or their *Figure*, or both.

Theoreme 3.

If Brickes moulded, or Stones squared *Cuneatim* (that is, *Wedge-wise*, broader above then below) shall be layd in a *Row levell*, with their ends supported, as in the precedent *Theoreme*, pointing all to one Center; then none of the pieces betweene can sinke till the *Supporters* give way, because they want roome in that *Figuration*, to descend *perpendicularly*. But this is yet a weake piece of *Structure*, because the *Supporters* are subject to much impulsion, especially if the line be long; for which reason this *Forme* is seldom used, but over *Windowes*, or narrowe *Doores*. Therfore to fortifie the *Work* as in this third *Theoreme* wee have supposed the *Figure* of all the *Materials* different from those in the second: So likewise wee must now change the *Posture*, as will appeare in the *Theoreme* following.

Theoreme 4.

If the *Materials* figured as before *Wedge-wise*, shall not be disposed levelly, but in forme of some Arch, or portion of a *Circle*, pointing all to the same *Center*: In this case neither the pieces of the sayd *Arch*, can sinke downewards, through want of roome to descend [*By the first Theor.*] perpendicularly: nor the *Supporters* or *Butments* (as they are tearmed) of the sayd *Arch* can suffer so much violence, as in the precedent flat Posture, for the roundnesse will alwayes make the Incumbent waight, rather to rest upon the *Supporters*, then to shove them; whence may be drawn an evident *Corolary*; that the safest of all *Arches* is the *Semicircular*, and of all *Vaults* the *Hemisphere*, though not absolutely exempted from some naturall weakenesse, [*which is the sole prerogative of perpendicular lines and right Angles*] as *Barnardino Baldi Abbot* of *Guastalla*, in his Commentary upon *Aristotles Mechaniques*, doth very well proove; where let me note by the way, that when any thing is *Mathematically* demostrated weake, it is much more *Mechanically* weake: Errors ever occurring more easily in the management of *Grosse Materials*, then *Lineall Designes*.

Theoreme 5.

As Semicircular *Arches*, or Hemispherical *Vaults*, being raised upon the totall *Diameter*, bee of all other the roundest, and consequently the securest, by the precedent *Theoreme*: So those are the gracefullest, which keeping precisely the same height, shall yet bee distended, one fourteenth part longer then the sayd entire *Diameter*; which addition of distent will conferre much to their *Beauty*, and detract but little from their *Strength*.

This observation I finde in *Leon-Batista Alberti*; But the practice how to preserve the same height, and yet distend the Armes or ends of the *Arch*, is in *Albert Durers Geometry*, who taught the *Italians* many an excellent *Line*, of great use in this *Art*.

Upon these five *Theoremes*, all the skill of *Arching* and *Vaulting* is grounded: As for those *Arches*, which our Artizans call of the third and fourth point; And the *Tuscan* writers *diterzo*, and *di quarto acuto*, because they alwayes concurre in an acute *Angle*, and doe spring from division of the *Diameter*, into three, foure, or more parts at pleasure; I say, such as these, both for the naturall imbecility of the sharpe *Angle* it selfe, and likewise for their very *Uncomelinesse*, ought to bee exiled from judicious eyes, and left to their first inventors, the *Gothes* or *Lumbards*, amongst other *Reliques* of that barbarous *Age*.

[*Apertions.*]

Thus of my first *Partition* of the parts of every *Fabrique*, into five Heads, having gone through the two former, and beene incidently carried into this last doctrine touching *Arches* and *Vaults.* The next now in order are the *Apertions*; under which tearme I doe comprehend *Doores*, *Windowes*, *Staire-cases*, *Chiminies*, or other *Conducts.* In short, all *Inlets* or *Outlets*; To which belong two generall *Cautions.*

First, That they bee as few in number, and as moderate in Dimension, as may possibly consist with other due respects: for in a word, all *Openings* are *Weaknings*.

Secondly, That they doe not approach too neere the *Angles* of the Wals; for it were indeed a most essentiall *Solecisme* to weaken that part, which must strengthen all the rest: A precept well recorded, but ill practised by the *Italians* themselves, perticularly at *Venice*, where I have observed diverse *Pergoli*, or *Meniana* (as *Vitruvius* seemeth to call them, which are certaine ballised out-standings to satisfie curiousity of sight) very dangerously set forth, upon the very point it selfe, of the *Murall Angle*.

Now, Albeit I make haste, to the casting and comparting of the whole *Worke*, (being indeede the very Definitive Summe of this *Art*, to distribute usefully and gracefully a well chosen *Plot*) yet I will first under their severall Heads, collect breifly some of the choisest notes, belonging to these particular *Overtures*.

Of *Doores* and *Windowes*.

These *In-lets* of *Men* and of *Light*, I couple together, because I find their due Dimensions, brought under one Rule, by *Leone Alberti* (a learned Searcher) who from the Schoole of *Pythagoras* (where it was a fundamentall *Maxime*, that the *Images* of all things are latent in *Numbers*) doth determine the comeliest Proportion, betweene breadths and heights; Reducing *Symmetrie* to

Symphonie, and the *harmonie* of *Sounde*, to a kinde of *harmonie* in *Sight*, after this manner: The two principall *Consonances*, that most ravish the Eare, are by consent of all Nature, the *fift*, and the *Octave*; whereof the first riseth radically, from the proportion, betweene *two* and *three*. The other from the double *Intervalle*, betweene *One* and *Two*, or betweene *Two* and *Foure etc.* Now if we shall transport these proportions, from Audible to visible *Objects*; and apply them as they shall fall fittest (the nature of the Place considered) Namely in some *Windowes*, and *Doores*, the *Symmetrie* of *Two* to *Three*, in their Breadth and Length; In others the double as aforesaid; There will indubitably result from either, a gracefull and *harmonious* contentment, to the Eye; Which speculation though it may appeare unto vulgar *Artizans*, perhaps too subtile, and too sublime, yet wee must remember, that *Vitruvius* himselfe doth determine many things in his profession; by *Musicall* grounds, and much commendeth in an *Architect*, a *Philosophical* Spirit; that is, he would have him (as I conceave it) to be no superficiall, and floating *Artificer*; but a *Diver* into *Causes*, and into the *Mysteries* of *Proportion*; Of the *Ornaments*, belonging both to *Doores* and *Windowes*, I shall speake in other place; But let mee heere adde one observation; That our *Master* (as appeareth by divers passages, and particularly *lib*. 6. *cap*. 9.) seemes to have beene an extreame Lover of *Luminous Roomes*; And indeede I must confesse that a Franke Light, can misbecome noe *AEdifice* whatsoever, *Temples* onely excepted; which were anciently darke, as they are likewise at this day in some Proportion. *Devotion more requiring collected then defused Spirits*. Yet on the other side we must take heede to make a House (though but for civill use) all *Eyes*, like *Argus*; which in Northerne *Climes* would be too could, In *Southerne*, too hot: And therefore the matter indeede importeth more then a merry comparison. Besides, There is no part of *Structure* either more expencefull, then *Windowes*; or more ruinous; not onely for that vulgar reason, as being exposed to all violence of weather; but because consisting of so different and unsociable pieces, as *Wood*, *Iron*, *Leade*, and *Glasse*, and those small and weake, they are easily shaken; I must likewise remember one thing, (though it be but a *Grammatical* note) touching *Doores*. Some were *Fores*, and Some were *Valuæ*. Those (as the very word may seeme to import) did open *outwards*, These *inwards*; And were commonly of two *Leaves* or *Panes*, (as we call them) thereby requiring indeed, a lesser *Circuit* in their *unfoulding*; And therefore much in use among *Italians* at this day; But I must charge them with an Imperfection, for though they let in as well as the former, yet they keepe out worse.

Of Staire-cases.

To make a compleate *Staire-case*, is a curious peece of *Architecture*: The vulgar Cautions are these.

That it have a very liberall *Light*, against all Casualtie of *Slippes*, and *Falles*.

That the space above the *Head*, bee large and *Airy*, which the *Italians* use to call *Un bel-sfogolo*, as it were good *Ventilation*, because a man doth spend much breath in mounting.

That the *Halfe-paces* bee well distributed, at competent distances, for reposing on the way.

That to avoyd *Encounters*, and besides to gratifie the beholder, the whole *Staire-case* have no nigard *Latitude*, that is, for the principall *Ascent*, at least ten foot in *Royall Buildings*.

That the breadth of every single *Step* or *Staire* bee never lesse then one foote, nor more then eighteen inches.

That they exceede by no meanes halfe a foot in their height or thicknesse; for our *Legges* doe labour more in *Elevation*, then in *Distention*: These I say are familiar remembrances, to which let me adde,

That the steps bee layd where they joyne *Con un*

tantino di scarpa; we may translate it somewhat *sloaping*, that so the foot may in a sort both ascend and descend together, which though observed by few, is a secret and delicate deception of the paines in mounting.

Lastly, to reduce this doctrine to some *Naturall*, or at least *Mathematicall* ground, (our *Master*, as we see, *lib.* 9. *cap.* 2.) borroweth those proportions, that make the sides of a *Rectangular Triangle*, which the Ancient Schoole did expresse in lowest tearmes, by the numbers of 3. 4. and 5. That is, *Three* for the *Perpendicular*, from the *Staire-head* to the ground: *Foure* for the *Ground-line* it selfe, or *Recession* from the wall; And *Five* for the whole *Inclination* or slopenesse in the ascent, which proportion, saith he, will make *Temperatas graduum librationes.* Hitherto of *Staire-cases* which are direct: There are likewise *Spirall*, or *Cockle staires*, either *Circular*, or *Ovall*, and sometimes running about a *Pillar*, sometimes vacant, wherein *Palladio*, (A man in this point of singular felicity) was wont to divide the *Diameter*, of the first sort into three parts, yeelding one to the *Pillar*, and two to the *Steps*; Of the second in to foure, whereof he have two to the *Staires*, and two to the *Vacuitie*, which had all their light from above, And this in exact *Ovals*, is a Master-piece.

Of *Chiminies.*

In the present businesse, *Italians* (who make very frugall fires, are perchance not the best Counsellers.) Therefore from them we may better learne, both how to raise faire *Mantels* within the roomes, and how to disguise gracefully the shafts of *Chimnies* abroad (as they use) in sundry formes (which I shall handle in the latter part of my labour) and the rest I well extract from *Philippe de l'Orme*: In this part of his *Worke* more diligent, then in any other, or, to doe him right, then any man else.

First, hee observeth very soberly, that who in the disposition of any Building will consider the nature of the *Region*, and the *Windes* that ordinarily blow, from this, or that *Quarter*; might so cast the roomes, which shall most need fire; that hee should little feare the incommodity of *Smoake*, and therefore hee thinkes, that inconvenience, for the most part to proceede from some inconsiderate beginning. Or if the error lay not in the *Disposition* but in the *Structure* it selfe; then hee makes a *Logicall* enquiry; That either the *Winde* is too much let in above, at the mouth of the *Shafte*, or the *Smoke* stifeled below; If none of these, Then there is a repulsion of the *Fume*, by some higher *Hill* or *Fabrique*, that shall overtoppe the *Chimney* and worke the former effect: If likewise not this, Then he concludes, that the Roome which is infested, must bee necessarily both little and close, so as the smoke cannot issue by a naturall *Principle*, wanting a succession and supply of new *Ayre*.

Now, In these cases he suggesteth divers Artificall remedies, of which I will allow one, a little *Description*, because it savoureth of *Philosophie*, and was touched by *Vitruvius* himselfe, *lib*. 1. *cap*. 6. but by this man ingeniously applied to the present use: Hee will have us provide to the present use: Hee will have us provide two hollow *brasse Balles* of reasonable capacitie, with little holes open in both, for reception of Water, when the Aire shalbe first sucked out; One of these wee must place with the hole upwards, upon an yron Wire, that shall traverse the *Chimney*, a little above the *Mantell*, at the ordinary height of the sharpest heate or flames, whereof the water within being rarified, and by rarifaction resolved into *Winde*, will breake out, and so force up the smoke, which otherwise might linger in the *Tunnell*, by the way, and oftentimes revert; With the other, (saith he) wee may supply the place of the former, when it is exhausted, or for a neede blow the *Fire* in the meane while; Which Invention I have interposed for some little intertainement of the Reader; I will conclude with a note from *Palladio*, who observeth that the Ancients did warme their Roomes, with certaine secret *Pipes* that came through the

Walles, transporting heate (as I conceive it) to sundry parts of the House, from one common *Furnace*; I am ready to baptize them *Caliducts*, as well as they are tearmed *Venti-ducts*, and *Aquæ-ducts* that convey Winde and Water; which whether it were a custome or a delicacie, was surely both for thrift, and for use, far beyond the German *Stoves*; And I should preferre it likewise before our owne fashion, if the very sight of a fire, did not adde to the Roome a kinde of *Reputation*, as old *Homer* doth teach us in a verse, sufficient to proove that himselfe was not blinde, as some would laie to his charge.

[*Conducts* for the *Suillage*.]

Touching *Conducts* for the *Suillage* and other necessities of the House, (which how base foever in use, yet for health of the Inhabitants, are as considerable, and perhaps more then the rest) I finde in our Authors, this Counsell; That *Art* should imitate *Nature*, in those ignoble conveyances; and separate them from Sight, (where there wants a running Water) into the most remote, and lowest, and thickest part of the *Foundation*: with secret vents passing up through the Walles like a *Tunnell* to the wilde Aire aloft: which all Italian Artizans commend for the discharge of noysome vapours, though else-where to my knowledge little practiced.

[*Compartition*: generall Precautions.]

Thus having considered the precedent *Appertions*, or *Overtures*, in severalitie according to their particular Requisites, I am now come to the casting and *Contexture* of the whole *Worke*, comprehended under the tearme of *Compartition*: Into which (being the mainest piece) I cannot enter without a few generall Precautions, as I have done in other Parts.

First therefore, Let no man that intendeth to build, setle his Fancie upon a draught of the *Worke* in *paper*, how exactly soever measured, or neately set off in *perspective*; And much lesse upon a bare *Plant* thereof, as they call the *Schiographia* or *Ground lines*; without a *Modell* or *Type* of the whole *Structure*, and of every parcell and Partition in *Pastboord* or *Wood*.

Next that the said Modell bee as plaine as may be, without colours or other beautifying, lest the pleasure of the *Eye* preoccupate the *Judgement*; which advise omited by the *Italian Architects*, I finde in *Philippe de l'Orme*, and therefore (though *France* bee not the *Theater* of best Buildings) it did merit some mention of his name.

Lastly, the bigger that this *Type* be, it is still the better, not that I will perswade a man to such an enormity, as that *Modell* made by *Antonio Labaco*, of Saint *Peters* Church in *Rome*, containing 22. foot in length, 16. in breadth, and 13. in heighth, and costing 4184. crownes: The price in truth of a reasonable *Chappell*: Yet in a *Fabrique* of some 40. or 50. thousand pounds charge, I wish 30. pounds at least layd out before hand in an exact *Modell*; for a little misery in the *Premises*, may easily breed some absurdity of greater charge, in the *Conclusion*.

[The *Compartition*.]

Now, after these premonishments, I will come to the *Compartition* it selfe; By which, the Authors of this *Art* (as hath beene touched before) doe understand, a *gracefull* and *usefull* distribution, of the whole *Ground plot* both for roomes of *Office*, and of *Reception* or *Entertainement*, as farre as the *Capacity* thereof, and the nature of the *Countrey* will comport. Which circumstances in the present *Subject*, are all of maine consideration, and might yeeld more discourse then an *Elementall Rapsodie* will premit. Therefore (to anatomize briefly this Definition) the *Gracefulnesse* (whereof wee speake) will consist in double *Analogie*, or correspondencie. First, between the *Parts* and the *Whole*, whereby a *great Fabrique* should have *great partitions*, *great Lights*, *great Entrances*, *great Pillars* or *Pylasters*; In summe, all the *Members great*. The next betweene the *Parts* themselves, not only, considering their *Breadths*, and *Lengths*, as before, when wee spake of *Doores* and *Windowes*; but here likewise enters a third respect of *Height*, a point (I must confesse) hardly reduceable to any generall precept.

True it is, that the *Ancients* did determine the *Longitude* of all *Roomes*, which were longer then broade, by the double of their *Latitude*, *Vitruvius lib*. 6. *cap*. 5. And the *Heighth* by the halfe of the breadth and length summed together. But when the Roome was precisely square, they made the *Height* half as much more as the *Latitude*; which Dimensions the moderne *Architects* have taken leave to varie upon discretion: Sometimes squaring the *Latitude*, and then making the *Diagoniall* or overthwart Line, from *Angle* to *Angle*, of the said *Square*, the measure of the *Heighth* sometimes more, but seldome lower then the full breadth it selfe; which boldnesse of quitting the old *Proportions*, some attribute first to *Michael Angelo da Buonaroti*, perchance upon the credite he had before gotten, in two other *Arts*.

[*Usefulnesse.*]

The second point is *Usefulnesse*, which will consist in a sufficient *Number* of Roomes, of all sorts, and in their apt *Coherence*, without *distraction*, without *confusion*; so as the beholder may not onely call it, *Una Fabrica ben raccolta*: as *Italians* use to speake of well united Workes, but likewise that it may appeare *airie* and *spiritous*, and fit for the welcome of cheerefull Guests; about which the principall difficultie will bee in contriving the *Lightes*, and *Staire-cases*, whereof I will touch a note or two: For the first, I observe that the ancient *Archietects* were at much ease. For both the *Greekes* and *Romanes* (of whose private dwellings *Vitruvius* hath left us some description) had commonly two *Cloystered* open Courts, one serving for the *Womens* side, and the other for the *Men*: who yet perchance nowadayes would take so much seperation unkindly. Howsoever, by this meanes, the reception of *light* into the Bodie of the building, was very prompt, both from without and from within: which we must now supplie either by some open *Forme* of the *Fabrique*, or among gracefull refuges, by *Tarrasing* any *Storie*, which is in danger of darkenesse; or lastly, by *perpendicular lights*, from the *Roofe*: of all other the most naturall, as shalbe shewed anon. For the second dificultie: which is casting of the *Stayre-cases*; That being in it selfe no hard point, but onely as they are incombrances of roome for other use: (which lights were not) I am therefore aptly moved heere to speake of them. And first of Offices.

[The *Offices*.]

I have marked a willingnesse, in the *Italian* Artisans, to destribute the *Kychin*, *Pantrie*, *Bakehouse*, *washing Roomes*; and even the *Buttrie* likewise, under ground; next above the *Foundation*, and sometimes Level with the plaine, or *Floore* of the *Cellar*: raysing the first *Ascent* into the house *Fifteene Foote* or more for that Ende, which besides the benefit of removing such *Annoyes* out of sight, and the gayning of so much more roome above, doth also by elevation of the *Front*, adde *Majestie* to the whole *Aspect*. And with such a disposition of the principall *Stayre-case*, which commonly doth deliver us, into the *Plaine* of the second *Storie*, there may bee wonders done, with a little roome, whereof I could alleadge brave *Examples* abroad; and none more Artificiall, and Delicious, then a House built by *Daniele Barbaro Patriarche* of *Aquileia* before mentioned, among the memorable *Commenters* upon *Vitruvius*. But the *Definition* (above determined) doth call us to some consideration of our owne Countrie, where though all the other pettie *Offices* (before rehearsed) may well enough bee so remote, yet by the naturall *Hospitalitie* of *England*, the *Buttrie* must be more visible, and wee neede perchance for our *Raunges*, a more spacious and luminous *Kitchin*, then the foresaid *Compartition* will beare; with a more competent neerenesse likewise to the *Dyning Roome* Or else besides other Inconveniences, perhapes some of the Dishes may straggle by the way; heere let me note a common defect, that wee have of a very usefull *Roome*, called by the *Italians Il Tinello*; and famílliar, nay almost essentiall, in all their great *Familes*. It is a Place properly appointed, to conserve the meate that is taken from the Table, till the *Waiters* eate, which with us by an olde fashion, is more unseemely set by, in the meane while.

[*Lodging* chambers.]

Now touching the distribution of *Lodging* chambers; I must here take leave to reproove a fashion, which I know not how hath prevailed through *Italie*, though without ancient examples, as farre as I can perceive by *Vitruvius*. The thing I meane, is, that they so cast their *partitions* as when all *Doors* are open, a man may see through the whole *House*; which doth necessariely put an intollerable servitude upon all the *Chambers* save the *Inmost*, where none can arrive, but through the rest; or else the *Walles* must be extreame thicke for secret passages. And yet this also will not serve the turne, without at least *Three* doores to every Roome: A thing most insufferable, in cold and windie Regions, and every where noe small weakening to the whole *Worke*; Therefore with us that want no cooling, I cannot commend the direct opposition of such *Overtures*, being indeede meerely grounded upon the fond ambition of displaying to a *Stranger* all our *Furniture* at one *Sight*, which therefore is most maintained by them that meane to harbour but a few; whereby they make onely advantage of the *vanitie*, and seldome prove the *Inconvenience*. There is likewise another defect (as absurdities are seldome solitarie) which will necessarily follow, upon such a servile disposing of inward Chambers. That they must bee forced to make as many common great *Roomes*, as there shalbe severall *Stories*; which (besides that they are usually darke, a point hardly avoided, running as they doe, through the middle of the whole *House*) doe likewise devoure so much Place, that thereby they want other *Galleries*, and Roomes of *Retreate*, which I have often considered among them (I must confesse) with no small wonder; for I observe no *Nation* in the World, by *Nature* more *private* and *reserved*, then the *Italian*, and on the other side, in no *habitations* lesse *privacie*; so as there is a kinde of Conflict, betweene their *Dwelling*, and their *Being*: It might heere perchance bee expected, that I should at least describe (which others have done in *draughts* and *designes*) divers *Formes* of *Plants* and *Partitions*, and varieties of *Inventions*; But speculative *Writers* (as I am) are not bound, to comprise all particular Cases, within the Latitude of the *Subject*, which they handle; Generall Lights, and *Directions*, and pointings at some faults, is sufficient. The rest must be committed to the sagacitie of the *Architect*, who will bee often put to divers ingenious shifts, when hee is to wrestle with scarsitie of *Ground*. As sometimes [The *Italians* call it *une stanze damnata*, as when a Buttrie is cast under a stayre-Case, or the like] to damme one Roome (though of speciall use) for the benefit and beautie of all the rest; Another while, to make those fairest, which are most in *Sight*, and to leave the other (like a cunning Painter) in shadow, *cum multis alijs*, which it were infinite to pursue. I will therefore close this Part touching *Compartition*, as cheerfully as I can with a short description of a *Feasting* or *entertaying Roome*, after the *Ægyptian* manner, who seeme (at least till the time of *Vitruvius*) from the auncient *Hebrewes* and *Phenicians* (whence all knowledge did flow) to have retayned, with other *Sciences*, in a high degree, also the *Principles*, and *practise* of this magnificent *Art*. For as farre as I may conjecture by our Masters Text, *lib*. 6. *cap*. 5. (where as in many other Places he hath tortured his

Interpreters) there could no *Forme*, for such a Royall use, bee comparably imagined, like that of the foresaide *Nation*, which I shall adventure to explaine.

[*Floore* or *Area.*]

Let us conceive a *Floore* or *Area* of goodly length, (For example, at least of 120 foote) with the *breadth* somewhat more then halfe of the *Longitude*, whereof the reason shalbe afterwards rendred. About the two longest *sides*, and *Head* of the said *Roome*, shall runne an Order of *Pillars*, which *Palladio* doth suppose *Corinthian* (as I see by his designe) supplying that point out of *Greece*, because we know no *Order*, proper to *Ægypt*. The fourth *side* I wil leave free for the *Entrance*: On the foresaid *Pillars* was laid an *Architrave*, which *Vitruvius* mentioneth alone: *Palladio* addes thereunto (and with reason) both *Freeze* and *Cornice*, over which went up a continued *Wall*, and therein, halfe or three quarter *Pillars*, answering directly, to the Order below, but a fourth Part lesse, and betweene these half *Columnes* above, the whole Roome was windowed around about.

Now, from the lowest *Pillars* there was layd over a *Contignation* of Floore, borne upon the outward *Wall*, and the *Head* of the *Columnes* with *Tarrace* and *Pavement*, *Sub dio* (saith our Master) and so indeed hee might safely determine the matter in *Egypt*, where they feare no *Clowds*: Therefore *Palladio* (who leaveth this *Tarrace* uncovered in the middle, and *ballised* about) did perchance construe him rightly, though therein discording from others: Alwayes we must understand a sufficient breadth of *Pavement*, left between the open part and the *Windowes*, for some delight of *Spectators*, that might looke downe into the *Roome*: The *Latitude* I have supposed contrary to some former Positions, a little more then the halfe of the length; because the Pillars standing at a competent distance from the outmost *Wall*, will by interception of the *Sight*, somewhat in appearance diminish the breadth; In which cases (as I have touched once or twice before) *Discretion* may bee more licentious then *Art*. This is the description of an *Egyptian* roome, for *Feastes* and other *Jollities*. About the *Walles* whereof wee must imagine entire *Statues*, placed below, and illuminated by the descending Light, from the *Tarrace*, as likewise from the *Windowes* betweene the halfe *Pillars* above: So as this *Roome* had abundant and advantageous *Light*; and besides other garnishing, must needes receive much *State* by the very heighth of the *Roofe*, that lay over two orders of *Columnes*: And so having runne through the foure parts of my first generall *Division*, namely, *Foundation*, *Walles*, *Appertions*, and *Compartition*; the *House* may now have leave to put on his *Hatte*: having hitherto beene uncovered it selfe, and consequently unfit to cover others. Which point though it be the last of this *Art* in execution, yet it is alwayes in *Intention* the first, For who would build but for *Shelter*? Therefore obtayning both the *Place*, and the dignity of a *Small cause*, it hath beene diligently handled by diverse, but by none more learnedly then *Bernardino Baldi* Abbot of *Guastalla* (before cited upon other occasion) who doth fundamentally, and *Mathematically* demonstrate the firmest *Knittings* of the upper *Timbers*, which make the *Roofe*. But it hath beene rather my *Scope*, in these *Elements* to fetch the ground of all, from *Nature* her selfe, which indeed is the simplest mother of *Art*. Therefore I will now onely deliver a few of the properest, and (as I may say) of the naturalest considerations, that belong to this remaining *Piece*.

[*Cover* or *Roofe*.]

There are two extremities to be avoyded in the *Cover*, or *Roofe*: That it be not too *heavy*, nor too *light*. The first, will suffer a vulgar objection of pressing too much the under-worke. The other, contayneth a more secret inconvenience; for the *Cover* is not onely a bare defence, but likewise a kinde of *Band* or *Ligature*, to the whole

Fabrique, and therefore would require some reasonable weight. But of the two extreames, a House *Top-heavie* is the worst. Next there must bee a care of *Equality*, that the *Edifice* be not pressed on the one side more then on the other; and here *Palladio* doth wish (like a cautelous *Artizan*) that the inward *Walles* might beare some good share in the burthen, and the outward be the lesse charged.

Thirdly, the *Italians* are very precise in giving the *Cover* a gracefull *pendence* or *slopenesse*, dividing the whole breadth into *Nine* parts; whereof two shall serve for the elevation of the highest *Toppe* or *Ridge*, from the lowest. But in this point the quality of the *Region* is considerable: For (as our *Vitruvius* insinuateth) those *Climes* that feare the falling and lying of much *Snow*, ought to provide more inclining *Pentices*: And *Comelinesse* must yeeld to *Necessity*.

These are the usefullest Cautions which I finde in *Authors*, touching the last *Head* of our Division, wherewith I will conclude the first *Part* of my present *Travaile*. The second remayneth, concerning *Ornaments* within, or without the *Fabrique*: A Piece not so dry as the meere Contemplation of proportions. And therefore I hope therein, somewhat to refresh both the *Reader*, and my selfe.

O F
T H E E L E M E N T S
O F
A R C H I T E C T U R E.

The II. part.

[*Ornaments* both *within* and *without* the *Fabrique.*]

EVERY Mans proper *Mansion* House and Home, being the *Theater* of his *Hospitality*, the *Seate* of *Selfe-fruition*, the *Comfortablest part* of his owne *Life*, the *Noblest* of his Sonnes *Inheritance*, a kinde of private *Princedome*; Nay, to the *Possessors* thereof, an *Epitomie* of the whole *World*: may well deserve by these *Attributes*, according to the degree of the *Master*, to be *decently* and *delightfully* adorned. For which ende, there are two *Arts* attending on *Architecture*, like two of her principall *Gentlewomen*, to *dresse* and *trimme* their *Mistresse*; PICTURE and SCULPTURE: Between whom, before I proceed any further, I wil venture to determine an ancient quarrell about their *Precedency*, with this *Distinction*; that in the garnishing of *Fabriques*, *Sculpture* no doubt must have the *preheminence*, as being indeede of neerer affinity to *Architecture* it selfe, and consequently the more naturall, and more sutable *Ornament*. But on the other side, (to consider these two *Arts* as I shall doe *Philosophically*, and not *Mechanichally*) An excellent Piece of *Painting*, is to my judgement the more admirable *Object*, because it comes neere an *Artificiall Miracle*; to make diverse distinct *Eminences* appeare upon a *Flat*, by force of *Shadowes*, and yet the *Shadowes* themselves not to appeare: which I conceive to be the uttermost value and vertue of a *Painter*, and to which very few have arrived in all *Ages*.

In these two *Arts* (as they are appliable to the *Subject* which I handle) it shall bee fit first to consider how to *choose* them; and next, how to *dispose* them. To guide us in the *choyce*, wee have a *Rule* somewhere (I well remember) in *Pliny*, and it is a prettie observation: That they doe mutually helpe to censure one another. For *Picture* is best when it standeth off, as if it were carved; and *Sculpture* is best when it appeareth so *tender*, as if it were *painted*, I meane, when there is such a seeming softnesse in the *Limbes*, as if not a *Chissell* had hewed them out of *Stone*, or other *Materiall*, but a *Pensill* had drawne and stroaked them in *Oyle*, which the judicious *Poet* tooke well to his Fancy.

Excudent alij spirantiamollius æra.

But this generalitie, is not sufficient to make a good *chooser*, without a more particular contraction of his Judgement. Therefore when a Piece of *Art*, is set before us, let the first Caution be, not to aske who made it, least the *Fame* of the *Author* doe Captivate the *Fancie* of the *Buyer*. For, that excellent *Men* doe alwaies *excellently*, is a false Conclusion; whereupon I observe among *Italian* Artizans three notable *Phrases*, which well decipher the degrees of their *Workes*.

They will tell you, that a thing was done *Con diligenza*, *Con studio*, and *Con Amore*; The first, is but a bare and *ordinary diligence*, The second, is a *learned diligence*; The third is much more, even a *loving diligence*; They meane not with love to the *Bespeaker* of the *Worke*, but with a love and delight in the *Worke* it selfe, upon some speciall *Fancie* to this, or that *Storie*; And when all these concurre (perticularly the last) in an eminent *Author*, Then perchance *Titianus Fecit*, will serve the *Turne*, without farther *Inquisition*; Otherwise Artizans have not onely their *growthes* and *Perfections*, but likewise their *Vaines* and *Tymes*.

The next Caution must be (to proceede *Logically*) that

in Judging of the *Worke* it selfe, wee bee not distracted with too many things at once; Therefore first (to beginne with *Picture*) we are to observe whether it bee well *drawne*, (or as more elegant *Artizans* tearme it) well *Design'd*; Then whether it be well *Coloured*, which bee the two generall *Heads*; And each of them hath two principall *Requisites*; For in well *Designing*, there must bee *Truth* and *Grace*, In well *Colouring*, *Force*, and *Affection*; All other *Praises*, are but Consequences of these.

Truth (as we Metaphorically take it in this *Art*) is a *Just* and *Naturall Proportion*, in every Part of the determined *Figure*. *Grace* is a certaine *free disposition*, in the whole *Draught*, answerable to that unaffected *franknes* of *Fashion*, in a living *Bodie*, Man or Woman, which doth *animate Beautie* where it is, and *supplie* it, where it is not.

Force consisteth, in the *Roundings* and *Raisings* of the *Worke*, according as the *Limbes* doe more or lesse require it; So as the *Beholder*, shall spie no *sharpenesse* in the bordering *Lines*; As when Taylors cut out a Sute, which *Italians* doe aptly tearme according to that comparison, *Contorni taglienti*; Nor any *flatnesse* within the *Bodie* of the *Figure*, which how it is done, we must fetch from a higher *Discipline*; For the *Opticques* teach us. That a *plaine*, will appeare *prominent*, and (as it were) *embossed*, if the Parts farthest from the *Axel-tree* or middle *Beame* of the *Eye*, shalbe the most *shadowed*. Because in all *Darknesse*, there is a kinde of *Deepenesse*, But as in the *Arte* of *perswasion*, one of the most Fundamentall *Precepts* is; the *concealement* of *Arte*, Soe here likewise, the *Sight* must be sweetly deceaved, by an insensible passage, from *brighter* colours, to *dimmer*, which *Italian Artizans* calle the *middle Tinctures*; That is, Not as the *whites*, and *yolkes* of *Egges* lie in the *Shell*, with visible distinction; But as when they are *beaten*, and *blended*, in a *Dish*: which is the neerest comparison, that I can suddenly conceive.

Lastly, *Affection* is the *Lively Representment*, of any *passion* whatsoever, as if the *Figures* stood not upon a *Cloth* or *Boorde*, but as if they were *acting* upon a *Stage*; And heere, I must remember, in truth with much marveile, a note, which I have received, from excellent *Artizans*, that though *Gladnesse*, and *Griefe*, be opposites in *Nature*; yet they are such *Neighbours* and *Confiners* in *Arte*, that the least touch of a *Pensill*, will translate a *Crying*, into a *Laughing* Face; which *Instance*, besides divers other, doth often reduce unto my memorie, that Ingenious *Speculation*, of the *Cardinall Cusanus* extant in his *Workes*, touching the *Coincidence* of *extreames*. And thus much of the foure *Requisites*, and *Perfections* in *Picture*.

In *Sculpture* likewise, the *Two* first are absolutely necessarie; The *third* impertinent; For *Solide* Figures neede not elevation, by force of *Lights*, or *shadowes*; Therefore in the Roome of this, wee may put (as hath beene before touched) a kinde of *Tendernesse*, by the *Italians* tearmed *Morbidezza*, wherein the *Chissell*, I must confesse, hath more glory then the *Pensill*; that being so hard an *Instrument*, and working upon so unpliant stuffe, can yet leave *Strokes* of so gentle *appearance*.

The *Fourth*, which is the expressing of *Affection* (as farre as it doth depend upon the *Activity*, and *Gesture* of the *Figure*) is as proper to the *Carver*, as to the *Painter*; though *Colours*, no doubt, have therein the greatest *Power*; whereupon, perchance, did first grow with us the *Fashion* of *colouring*, even *Regall Statues*, which I must take leave to call an *English Barbarisme*.

Now in these fower *Requisites* already rehearsed, it is strange to note, that no *Artizan*, having ever beene blamed for exceese in any of the three last; onely *Truth* (which should seeme the most *Innocent*) hath suffered some Objection, and all *Ages*, have yeelded some one or two *Artificers*, so *prodigiously* exquisite, that they have beene reputed too *Naturall*, in their *Draughts*; which will well appeare, by a famous Passage in *Quintilian*, touching the *Characters* of the ancient Artizans, falling now so aptly into my memory, that I must needes translate it, as in truth it may well deserve.

The Place which I intend, is extant in the last *Chapter* save one of his whole *Worke*, beginning thus in *Latine*.

Primi, quorum quidem opera non vetustatis modo gratia visenda sunt clari Pictores suisse dicuntur, Polygnotus atque Aglaophon, etc.

The whole Passage, in *English* standeth thus:

THE first *Painters* of name, whose *Workes* bee considerable for any thing more then onely *Antiquitie*, are said to have beene *Polygnotus*, and *Aglaophon*; whose bare *Colourings* (hee meanes I thinke in *white* and *blacke*) hath even yet so many followers, that those rude and first *Elements*, as it were of that, which within a While, became an *Arte*, are preferred, before the greatest *Painters* that have beene extant after them, out of a certaine *Competition* (as I conceive it) in point of *Judgement*. After these, *Zeuxes* and *Parasius* not farre distant in *age*, both about the time of the *Peloponesion Warre*, (for in *Xenophon* wee have a Dialogue betweene *Parasius* and *Socrates*) did adde much to this *Arte*. Of which the *first* is said, to have invented the due disposition of *Lights* and *Shadowes*; The *second*, to have more subtilly examined, the *truth* of *Lines* in the Draught; for *Zeuxes* did make *Limbes*, bigger then the *life*; deeming his *Figures*, thereby the more *stately* and *Majesticall*; and therein (as some thinke) imitating *Homer*, whom the *stoutest* forme doth please, even in *Women*. On the other side, *Parasius* did exactly limit al the *Proportions* so, as they call him the *Law-giver*, because in the *Images* of the *Gods* and of *Heroicall Personages*, others have followed his *Paternes* like a *Decree*; But *Picture* did most flourish, about the daies of *Phillip* and even to the Successours of

Alexander; yet by sundry *habilities*; for *Protogenes*, did excell in *Diligence*; *Pamphilus* and *Melanthius* in due *Proportion*, *Antilphilus* in a *Franke Facilitie*; *Theon* of *Samos*, in strength of *Fantasie* and conceiving of *Passions*; *Apelles*, in *Invention*, and *Grace*, whereof he doth himselfe most vaunte; *Euphranor*, deserves admiration, that being in other excellent *studies*, a principall Man, he was likewise a wondrous Artizan, both in *Painting* and *Sculpture*. The like difference we may observe among the *Statuaries*; for the workes of *Calon* and *Fgesias* were somewhat *stiffe*, like the *Tuscan* manner; Those of *Calamis* not done with so *cold* stroakes; And *Myron* more *tender* then the former; a diligent *Decency* in *Polycletus* above others, to whom though the highest prayse bee attributed by the most, yet lest he should goe free from exception, some thinke hee wanted *solemnesse*; for as he may perchance be sayd to have added a *comely dimension* to humane shape, somewhat above the *truth*; so on the other side, hee seemed not to have fully expressed the *Majesty* of the *Gods*: Moreover, hee is sayd not to have medled willingly with the *graver age*, as not adventuring beyond *smooth cheekes*: But these vertues that were wanting in *Polycletus*, were supplied by *Phidias* and *Alcmenes*, yet *Phidias* was a better Artizan in the representing of *God*, then of *Men*; and in his workes of *Ivorie*, beyond all emulation, even though hee had left nothing behinde him, but his *Minerva* at *Athens*, or the *Olympian Jupiter* in *Elis*, whose *Beautie* seemes to have added somewhat, even to the received *Religion*; the *Majestie* of the *Worke*, as it were equalling the *Deity*. To *Truth*, they affirme *Lysippus* and *Praxiteles*, to have made the neerest approach: for *Demetrius* is therein reprehended, as rather exceeding then deficient; having beene a greater aymer at *Likenes*, then at *Lovelines*.

This is that witty Censure of the ancient Artizans, which *Quintilian* hath left us, where the last *Charactar* of *Demetrius* doth require a little *Philosophicall* examination; How an *Artificer*, whose end is the *imitation* of *Nature*, can bee too *naturall*; which likewise in our dayes was either the fault, or (to speake more gently) the too much perfection of *Albert Durer*, and perhaps also of *Michael Angelo da Buonaroti*, betweene whom I have heard noted by an ingenious *Artizan*, a pretty nice difference, that the *German* did too much expresse *that which was*; and the *Italian*, *that which should be*: Which severe observation of *Nature*, by the one in her *commonest*, and by the other in her *absolutest* Formes, must needs produce in both a kinde of *Rigidity*, and consequently more *naturalnesse* then *gracefulnesse*: This is the cleerest reason, why some exact *Symmetrists* have been blamed, for being too true, as neere as I can deliver my conceit. And so much touching the choyce of *Picture* and *Sculpture*: The next is, the *application* of both, to the *beautifying* of *Fabriques*.

[*Picture* and *Sculpture* in the *beautifying* of *Fabriques*.]

First therefore touching *Picture*, there doth occurre a very pertinent doubt, which hath beene passed over too slightly, not onely by some *Men*, but by some *Nations*; namely, whether this *Ornament* can wel become the *Outside* of *houses*, wherein the *Germanes* have made so little scruple, that their best *Townes* are the most painted, as *Augusta* and *Norembergh*. To determine this question in a word: It is true, that a Story well set out with a good *Hand*, will every where take a *Judicious* eye: But yet withal it is as true, that various colours on the *Out-walles* of *Buildings*, have alwayes in them more *Delight* then *Dignity*: Therfore I would there admit no *Paintings* but in *Blacke* and *White*, nor even in that kinde any Figures (if the roome be capable) under *Nine* or *Ten* foot high, which will require no ordinary *Artizan*; because the faults are more *visible* then in small *Designes*. In unfigured *paintings* the noblest is, the imitation of *Marbles*, and of *Architecture* it selfe, as *Arches*, *Treezes*, *Columnes*, and the like.

Now for the *Inside*, heere growes another doubt, whether *Grotesca* (as the *Italians*) or *Antique* worke (as wee call it) should be received, against the expresse authorities of *Vitruvius* himselfe, *lib. 7. cap.* 5. where *Pictura* (saith hee) *Fit eius*, *quod est*, *seu potest esse*, excluding by this severe definition, all *Figures* composed of different *Natures* or *Sexes*; so as a *Syrene* or a *Centaure* had beene intolerable in his eye: But in this wee must take leave to depart from our *Master*, and the rather because he spake out of his owne *profession*, allowing *Painters* (who have ever bin as little limited as *Poets*) a lesse scope in their imaginations, even then the gravest *Philosophers*, who sometimes doe serve themselves of *Instances*, that have no *Existence* in *Nature*; as wee see in *Platoes Amphisbœna*, and *Aristotles Hirco-Cervus*. And (to settle this point) what was indeede more common and familiar among the *Romanes* themselves, then the *Picture* and *Statue* of *Terminus*, even one of their *Deities*? which yet if we well consider, is but a piece of *Grotesca*; I am for these reasons unwilling to impoverish that *Art*, though I could wish such *medlie* and *motlie* Designes, confined onely to the *Ornament* of *Freezes*, and Borders, their properest place. As for other *Storied* Workes upon *Walles*, I doubt our *Clime* bee too *yeelding* and *moist*, for such *Garnishment*; therefore leaving it to the *Dwellers* discretion, according to the qualitie of his *Seat*; I will onely adde a caution or two, about the disposing of *Pictures* within.

[The disposing of *Pictures* within.]

First, that no *Roome* bee furnished with too many, which in truth were a *Surfet* of *Ornament*, unlesse they bee *Galleries*, or some peculiar *Repository* for *Rarities* of *Art*.

Next, that the best *Pieces* be placed not where there is the *least*, but where there are the *fewest lights*; therefore not onely *Roomes* windowed on both ends, which we call through-lighted; but with two or more *Windowes* on the same side, are enemies to this *Art*; and sure it is, that no *Painting* can be seene in full *Perfection* but (as all *Nature* is illuminated) by a *single* Light.

Thirdly, that in the *placing* there be some care also taken, how the *Painter* did stand in the *Working*, which an intelligent *Eye*, will easily discover, and that *Posture* is the most naturall; so as *Italian* pieces will appeare best in a *Roome* where the *Windowes* are high; because they are commonly made to a descending *Light*, which of all other doth set off mens *Faces* in their truest Spirit.

Lastly, that they bee as *properly* bestowed for their *quality*, as *fitly* for their *grace*: that is, *chearefull* Paintings in *Feasting* and *Banquetting* Roomes; *Graver Stories* in *Galleries*, *Land-schips*, and *Boscage*, and such *wilde* workes in open *Tarraces*, or in *Summer houses* (as we call them) and the like.

And thus much of *Picture*, which let mee close with this Note; that though my former Discourse may serve perchance for some reasonable *leading* in the choyce of such delights; yet let no man hope by such a *speculative* erudition, to discerne the *Masterly* and *Mysterious* touches of *Art*, but an Artizan himselfe; to whom therefore we must leave the *prerogative*, to censure the manner and handling, as hee himselfe must likewise leave some points, perchance of no lesse value to others; as for example, whether the *Story* be rightly *represented*, the *Figures* in true *action*, the *Persons suted* to their severall qualities, the *affections proper* and *strong*, and such like observations.

[*Sculpture.*]

Now for *Sculpture*, I must likewise begin with a *Controversie*, as before (falling into this Place) or let mee rather call it a very meere *Fancie*, strangely taken by *Palladio*, who having noted in an old *Arch* or two at *Verona*, some part of the *Materials* already cut in fine *Formes*, and some *unpolished*, doth conclude (according to his *Logicke*) upon this particular, that the *Auncients* did leave the outward *Face*, of their *Marbles* or *Free-stone*, without any *Sculpture*, till they were laid, and *Cimented* in the bodie of the *Building*; For which likewise hee findeth a reason (as many doe now and then very wittily, even before the thing it selfe be true) that the Materialls being left *rough* were more managable in the *Masons* hand, then if they had beene *smooth*; And that so the *sides* might bee laide together the more exactly; Which Conceit, once taken hee seemes to have farther imprinted, by marking in certaine Storied *Sculptures*, of oulde time how preciesly the parts and *Lines* of the *Figures* that passe from one Stone to another, doe meete; which hee thinkes could hardly fall out so right (forgetting while he speakes of *ancient* things, the auncient *Diligence*) unlesse they had beene cut, after the joyning of the *Materials*, But all these *Inducements*, cannot countervaile the sole *Inconvenience* of *shaking*, and *Disjoynting* the *Commissures* with so many Strokes of the *Chissell*, besides an *Incommodious* Working on *Scaffolds*; especially having no testimonie, to confirme it, that I have yet seene among the records of *Art*; Nay, it is indeede rather true, that they did *Square*, and *Carve*, and *Polish*, their *Stone* and *Marble* Workes, even in the very *Cave* of the *Quarrie*, before it was hardened by open Aire; But (to leave disputation) I will set downe a few *Positive* notes, for the placing of *Sculpture*; because the chusing hath beene handled before.

[The placing of *Sculpture*.]

That first of all, it bee not too *generall* and *abundant*, which would make a *House*, looke like a *Cabbinet*, and in this point *morall Philosophie* which tempereth *Fancies*, is the *Superintendent* of *Art*.

That especially, There bee a due moderation of this *Ornament* in the first approach; where our Authors doe more commend (I meane about the Principall *Entrance*) a

Dorique, then a *Corinthian* garnishment; So as if the great *Doore*, be *Arched*, with some brave *Head*, cut in fine *Stone* or *Marble* for the *Key* of the *Arch*, and two *Incumbent* Figures gracefully leaning upon it, towards one another, as if they meant to conferre; I should thinke this a sufficient entertainement, for the first *Reception*, of any *Judicious* Sight, which I could wish seconded, with two great standing *Statues* on each side of a paved way that shall leade up into the *Fabrique*, So as the *Beholder* at the first entrance, may passe his *Eye* betweene them.

That the *Nices*, if they containe *Figures* of *white* Stone or *Marble*, bee not coloured in their *Concavitie* too *blacke*, For though *Contraria juxta se posita magis illuceseunt* (by an olde Rule) yet it hath beene subtilly, and indeede truely noted that our *Sight*, is not well contented, with those sudden departments, from one extreame, to another; Therefore let them have, rather a *Duskish* Tincture, then an absolute *blacke*.

That *fine* and delicate *Sculptures*, be helped with *Neerenes*, and *Grosse* with *distance*; which was well seene in the olde controversie, betweene *Phidias* and *Alcmenes* about the *Statue* of *Venus*: wherein the First did shew discretion, and save labour, because the *Worke* was to bee viewed at good *Height*, which did drowne the sweete and diligent strokes of his *Adversarie*: A famous emulation of two principall *Artizans*, celebrated even by the *Greeke Poets*.

That in the placing of standing *Figures* aloft, wee must set them in a *Posture* somewhat bowing forward; because (saith our Master, *lib*. 3. *cap*. 3.) out of a better *Art* then his owne) the *visuall beame* of our eye, extended to the *Head* of the said *Figures*, being longer then to the *Foote*, must necessarily make that *part* appeare *farther*; so as to reduce it to an erect or upright *position*, there must be allowed a due advantage of *stooping* towards us; which *Albert Durer* hath exactly taught, in his fore mentioned *Geometry*. Our *Vitruvius* calleth this affection in the *Eye*, a *resupination* of the *Figure*: For which word (being in truth his owne, for ought I know) wee are almost as much beholding to him, as for the observation it selfe: And let thus much summarily suffice, touching the choice and use of these adorning *Arts*. For to speake of garnishing the *Fabrique* with a *Row* of erected *Statues*, about the *Cornice* of every *Contignation* or Story, were discourse more proper for *Athens* or *Rome*, in the time of their true greatnesse, when (as *Plinie* recordeth of his own *Age*) there were neere as many carved *Images*, as living *Men*, like a noble contention, even in point of *Fertility*, betweene *Art* and *Nature*, which passage doth not onely argue an infinite abundance, both of *Artizans* and *Materials*; but likewise of *Magnificent* and *Majesticall* desires, in every common person of those times; more or lesse according to their *Fortunes*. And true it is indeed that the marble *Monuments* and *Memories* of well deserving Men, wherewith the very high wayes were *strewed* on each side was not a bare and transitory entertainement of the *Eye*, or onely a gentle deception of *Time*, to the *Travailer*: But had also a secret and strong *Influence*, even into the advancement of the *Monarchie*, by continuall representaion of vertuous examples; so as in that point *ART* became a piece of *State*.

[*Mosaique* and *Plastique*.]

Now as I have before subordinated *Picture*, and *Sculpture* to *Architecture*, as their Mistresse; so there are certaine inferiour ARTS likewise subordinate to them: As under Picture, *Mosaique*; under Sculpture, *Plastique*; which two, I onely nominate, as the fittest to garnish *Fabriques*.

Mosaique is a kinde of Painting in small *Pebbles*, *Cockles* and *Shells* of sundry colours; and of late dayes likewise with pieces of *Glasse*, figured at pleasure; an *Ornament* in trueth, of much beauty, and long life, but of most use in *pavements* and *Floorings*.

Plastique is not onely under *Sculpture*, but in deed

very *Sculpture* it selfe: but with this difference; that the *Plasterer* doth make his Figures by *Addition*, and the *Carver* by *Substraction*, whereupon *Michael Angelo* was wont to say somewhat pleasantly: That *Sculpture* was nothing but *A purgation of superfluities*. For take away from a piece of wood, or stone, all that is *superfluous*, and the remainder is the intended *Figure*. Of this *Plastique Art*, the chiefe use with us is in the gracefull *fretting* of roofes: but the *Italians* applie it, to the *manteling* of *Chimneys*, with great Figures. A cheape piece of *Magnificence*, and as durable almost within doores, as harder *Forms* in the weather. And here though it bee a little excursion, I cannot passe unremembred a gaine, their manner of disguising the shaftes of *Chimneys* in various fashions, whereof the noblest is the *Pyramidall*: beeing in trueth a piece of polite and civill discretion, to convert even the *conduits* of soote and smoake, into *Ornaments*; whereof I have hitherto spoken as farre as may concerne the Bodie of the Building.

[Garden *Ornaments*.]

Now there are *Ornaments* also without, as *Gardens*, *Fountaines*, *Groves*, *Conservatories* of rare *Beasts*, *Birds*, and *Fishes*. Of which ignobler kind of Creatures, *Wee ought not* (saith our greatest Master among the sonnes of Nature) *childishly to despise the Contemplation*; *for in all things that are naturall*, *there is ever something*, *that is admirable*. Of these externall delights, a word or two.

First, I must note a certaine contrarietie betweene *building* and *gardening*: For as Fabriques should bee *regular*, so Gardens should bee *irregular*, or at least cast into a very wilde *Regularitie*. To exemplifie my conceit; I have seene a *Garden* (for the maner perchance incomparable) into which the first Accesse was a high walke like a *Tarrace*, from whence might bee taken a generall view of the whole *Plott* below; but rather in a delightfull confusion, then with any plaine distinction of the pieces. From this the *Beholder* descending many steps, was afterwards conveyed againe, by severall *mountings* and *valings*, to various entertainements of his *sent*, and *sight*: which I shall not neede to describe (for that were poeticall) let me onely note this, that every one of these diversities, was as if hee had beene *Magically* transported into a new Garden.

But though other *Countreys* have more benefite of Sunne then wee, and thereby more properly tyed to contemplate this delight; yet have I seene in our *owne*, a delicate and diligent *curiositie*, surely without *parallel* among foreigne *Nations*: Namely, in the Garden of Sir *Henry Fanshaw*, at his seat in *Ware-Parke*, where I wel remember, hee did so precisely examine the *tinctures*, and *seasons* of his *flowres*, that in their *setting*, the *inwardest* of *those* which were to come up at the same time, should be alwayes a little *darker* then the *outmost*, and so serve them for a kinde of gentle *shadow*, like a piece not of *Nature*, but of *Arte*: which mention (incident to this place) I have willingly made of his *Name*, for the deare *friendship* that was long betweene us: though I must confesse, with much wrong to his other *vertues*; which deserve a more solide *memoriall*, then among these vacant observations. So much of *Gardens*.

[*Fountaines.*]

Fountaines are *figured*, or only plaine *Water'd-workes*: Of either of which, I will describe a matchlesse patterne.

The first, done by the famous hand of *Michael Angelo da Buonaroti*, in the figure of a sturdie *woman*, *washing* and *winding* of linnen clothes; in which Acte, shee *wrings* out the water that made the *Fountaine*; which was a gracefull and naturall conceit in the Artificer, implying this rule; That all *designes* of this kind, should be *proper*.

The other doth merite some larger expression; There went a long, straight, mossie walke of competent breadth, greene, and soft under foot, lifted on both sides with an

Aquæduct of white stone, breast-high, which had a hollow *channell* on the top, where ranne a prety trickling streame; on the *edge* whereof, were couched very thicke all along, certaine small *pipes* of lead, in little holes; so neatly, that they could not be well perceived, till by the turning of a cocke, they did sprout over interchangeably from side to side, above mans height, in forme of *Arches*, without any intersection or meeting aloft, because the pipes were not exactly opposite; so as the *Beholder*, besides that, which was fluent in the *Aquæducts* on both hands in his view, did walke as it were, under a continuall *bowre* or *Hemisphere* of water, without any drop falling on him. An *invention* for refreshment, surely farre excelling all the *Alexandrian* delicacies, and *Pneumatiques* of *Hero*.

[*Groves.*]

Groves, and artificall devices under ground, are of great expence, and little dignitie; which for my part I could wish converted here into those *Crypteria*, whereof mention is made among the curious provisions of *Ticho Braghe* the *Danish Ptolemie*, as I may well call him: which were deepe *concaves* in gardens, where the *starres* might be observed even at *noone*. For (by the way) to thinke that the brightnesse of the Sunnes body above, doth drowne our discerning of the lesser lights, is a popular errour; the sole impediment being that lustre, which by *reflection*, doth spread about us, from the face of the Earth; so as the *caves* before touched, may well conduce, not to a delicious, but to a learned pleasure.

[*Aviaries.*]

In *Aviaries* of wire, to keepe Birdes of all sorts, the *Italians* (though no wastfull Nation) doe in some places bestow vast expence; including great scope of *ground*, varietie of *bushes*, *trees* of good height, running *waters*, and sometimes a *Stove* annexed, to contemper the *Aire* in

Winter. So as those *Chanteresses*, unlesse they be such as perhaps delight as much in their wing, as in their voice, may live long, among so good provisions and roome, before they know that they are *prisoners*; reducing often to my memory, that conceit of the *Romane Stoicke*, who in comparison of his owne free *contemplations*, did thinke divers great and splendent fortunes of his time, little more then *commodious captivities*.

[*Ponds.*]

Concerning *Ponds* of pleasure neere the habitation; I will referre my selfe to a grave *Author* of our owne (though more illustrious by his other worke) namely *Sarisburiensis de Piscinâ.*

And here I will end the second part touching *Ornaments*, both *within*, and *without* the *Fabrique*.

[*Methode* for *Worke* to be *Commodious*, *Firme*, and *Delightfull.*]

Now as almost all those, which have delivered the *Elements* of *Logicke*, doe usually conclude, with a Chapter touching *Methode*; so I am heere seized with a kind of critical spirit, and desirous to shut up these building *Elements*, with some *Methodical* direction how to *censure* Fabriques alreadie raised: for indeed without some way to contract our *Judgement*, which among so many particulars would bee lost by diffusion; I should thinke it almost harder to be a good *Censurer*, then a good *Architect*: Because the *working* part may be helped with *Deliberation*, but the *Judging* must flow from an *extemporall habite*. Therfore, (not to leave this last *Piece* without some *Light*) I could wish him that commeth to examine any noble *Work*, first of all to examine himselfe, whether perchance the *sight* of many brave things before (which remaine like impressed formes) have not made him apt to thinke nothing *good*, but that which is the *best*; for this *humour* were too sowre. Next, before hee come to settle any imaginable opinion, let him by all meanes seeke to informe himselfe precisely, of the *Age* of the *Worke* upon which hee must passe his *Doome*. And if hee shall finde the apparent *decayes* to exceed the proportion of *Time*; then let him conclude without farther inquisition, as an absolute Decree, that either the *Materials* were too *slight*, or the *Seate* is nought. Now, after these premisses, if the *House* be found to beare his *yeares* well, (which is alwayes a token of sound *constitution*) Then let him suddenly runne backewardes, (for the *Methode* of *censuring* is contrary to the *Methode* of *composing*) from the *Ornaments* (which first allure the *Eye*) to the more essentiall *Members*, till at last hee be able to forme this *Conclusion*, that the *Worke* is *Commodious*, *Firme*, and *Delightfull*; which (as I said in the beginning) are the three capitall *Conditions* required in good *Buildings*, by all *Authors* both Ancient and Moderne. And this is, as I may tearme it the most *Scientificall* way of *Censuring*. There are two other which I must not forget. The first in *Georgio Vassari*, before his labourious *Worke* of the lives of *Architects*, which is to passe a running examination over the whole *Edifice*, according to the properties of a *well shapen Man*. As whether the *Wals* stand upright upon cleane footing and *Foundation*; whether the *Fabrique* bee of a beautifull *Stature*, whether for the *breadth* it appeare well *burnished*, whether the principall *Entrance* be on the middle Line of the *Front* or Face, like our *Mouthes*, whether the *Windowes*, as our *Eyes*, be set in equall number and distance on both sides, whether the *Offices* like the *Veines* in our Bodies, be usefully distributed, and so forth. For this *Allegoricall* review may be driven as farre as any *Wit* will, that is at leasure.

The second way, is in *Vitruvius* himselfe, *lib*. 1. *cap*. 2. where hee summarily determineth six *Considerations*, which accomplish this whole *Art*.

Ordinatio.
Dispositio.
Eurythmia.
Symmetria.
Decor, and
Distributio.

Whereof (in my conceit) wee may spare him the first two; for as farre as I can perceive, either by his *Interpreters*, or by his own *Text* (which in that very place, where perchance he should be *clearest*, is of all other the *Clowdiest*) hee meaneth nothing by *Ordination*, but a well setling of the *Modell* or *Scale* of the whole *Worke*. Nor by *Disposition*, more then a neate and full expression of the first *Idea* or *Designement* thereof; which perchance doe more belong to the *Artificer*, then to the *Censurer*. The other foure are enough to condemne, or absolve any *Fabrique* whatsoever. Whereof *Eurythmia* is that agreeable *Harmony*, betweene the breadth, length, and height of all the Roomes of the *Fabrique*, which suddenly where it is taketh every *Beholder*, by the secret power of *Proportion*: wherein let mee onely note this, That though the least error or offence that can be committed against *sight*, is excesse of height; yet that fault is no where of small importance, because it is the greatest offence against the Purse.

Symmetria is the *conveniencie* that runneth betweene the *Parts* and the *Whole*, whereof I have formerly spoken.

Decor is the keeping of a due *Respect* betweene the *Inhabitant*, and the *Habitation*. Whence *Palladius* did conclude, that the principall *Entrance* was never to be regulated by any certaine *Dimensions*; but by the dignity of the *Master*; yet to exceede rather, in the *more*, then in the *lesse*, is a marke of *Generosity*, and may alwayes be excused with some noble *Embleme*, or *Inscription*, as that of the *Conte di Beuilacqua*, over his large *Gate* at *Verona*, where perchance had beene committed a little *Disproportion*.

Patet Janua: *Cor magis*.

And heere likewise I must remember our ever memorable Sir *Philip Sidney*, (whose *Wit* was in truth the very rule of *Congruity*) who well knowing that *Basilius* (as hee had painted the *State* of his *Minde*) did rather want some extraordinary *Formes* to entertaine his *Fancie*, then *roome* for *Courtiers*; was contented to place him in a *Star-like Lodge*; which otherwise in severe *Judgement* of *Art* had beene an incommodious *Figure*.

Distributio is that usefull *Casting* of all Roomes for *Office*, *Entertainement*, or *Pleasure*, which I have handled before at more length, then any other Piece.

These are the Foure *Heads* which every man should runne over, before hee passe any determinate *Censure*, upon the Works that he shal view, wherewith I will close this last part, touching *Ornaments*. Against which (mee thinkes) I heare an *Objection*, even from some well-meaning man; That these delightfull *Craftes*, may be divers wayes ill applied in a *Land*. I must confesse indeede, there may bee a *Lascivious*, and there may be likewise a *superstitious* use, both of *Picture* and of *Sculpture*: To which possibility of misapplication, not onely these *Semi-liberall Arts* are subject; but even the highest perfections, and *endowments* of *Nature*. As *Beautie* in a light woman, *Eloquence* in a mutinous Man, *Resolution* in an Assasinate, Prudent *observation* of houres and humours, in a corrupt Courtier, *Sharpenesse* of wit and argument in a seducing Scholler; and the like. Nay, finally let mee aske, what ART can be more pernicious, then even RELIGION it selfe, if it selfe be converted into an Instrument of ART: Therefore, *Ab abuti ad non uti, negatur consequentia*.

Thus having stiched in some sort together, these *Animadversions*, touching *Architecture*, and the *Ornaments* thereof; I now feele that *contemplative* spirits are as restlesse as *active*; for doubting with my selfe, (as all weakenesse is jealous) that I may be thought to have spent my poore observation abroad, about nothing but Stone and Timber, and such Rubbage; I am thereby led into an immodestie of proclaiming another Worke, which I have long devoted to the service of my *Countrey*: Namely, A *Philosophicall Survey of Education*, which is indeed, a second *Building*, or repairing of Nature, and, as I may tearme it, a kinde of *Morall Architecture*; whereof such Notes as I have taken in my foreigne transcursions or abodes, I hope to utter without publike offence, though still with the freedome of a plaine *Kentish* man. In the meane while I have let these other *Gleanings* flie abroad, like the Bird out of the *Arke*, to discover what footing may bee, for that which shall follow.

FINIS.

4 John Harris and the *Lexicon Technicum*

John Harris: biographical note

John Harris was a clergyman-scholar born about 1666, the son of Edward Harris. He entered Trinity College, Oxford where he was scholar obtaining the degree of B.A. (1686) and M.A. (1689). After taking holy orders he secured a number of clerical appointments. In 1690 he had spiritual charge of the parish of Winchelsea and the following year he was inducted into the rectory of St. Thomas. At this period he was fortunate to be patronised by Sir William Cowper, later Baron of Wingham and Lord High Chancellor of Great Britain, to whom Harris subsequently dedicated Volume II of his *Lexicon Technicum*. Cowper appointed Harris as his personal chaplain and secured for him the post of prebend in the cathedral of Rochester (1707-8). Meanwhile Harris had further manifested his scholarly disposition by graduating B.D. at Cambridge University (1699) and obtaining the Lambeth degree of D.D. (1706).[1]

Harris possessed a first-class intellect with a disposition to science and a singular command of mathematics. Harris was, by way of illustration, one of the first interlocutors of Newton's revolutionary mathematical precepts. His achievements in science were recognised by his election to a Fellowship of the Royal Society (1696) and two years later he gave the Boyle lectures in St. Paul's Cathedral. About 1698 Harris used his mathematical abilities to give a course of public lectures 'for the public good' and, we can assume, to supplement his income: Harris, despite his considerable gifts and high connections, was culpably improvident and was generally in straitened circumstances. He continued to lecture on mathematics until 1704, the year in which Volume I of *Lexicon Technicum* was published and two years later he was elected a member of the council of the Royal Society, rising to the office of Secretary in 1709.

In addition to his mathematical aptitudes, John Harris possessed other qualities that were to serve him well in the creation of *Lexicon Technicum*: he was a punctilious editor-compiler. This facility was well recognised by the London book publishers and Harris produced an impressive sequence of works right up to the very end of his life in 1719. A summary listing of his principal publications gives some insight into

Harris's breadth of interests and places the *Lexicon Technicum* within the wider context of his work.

Harris's principal publications are: *Remarks on some late Papers relating to the Universal deluge, and to the Natural History of the Earth* (1697); *The Atheistical Objections against the Being of God, and his Attributes, fairly considered and fully refuted* (1698), this work is the text of the Boyle lectures; *Short but yet plain Elements of Geometry and Plane Trigonometry* (1701); *The description and uses of the Celestial and Terrestrial Globes, and Collins's Pocket Quadrant* 1703, this book is typical of many such works written in the 18th century by mathematical practitioners such as John Harris; *Lexicon Technicum* (1704-10); *Navigantium atque Itinerantium Bibliotheca: or a compleat Collection of Voyages and Travels, consisting of above four hundred of the most authentick writers* (1705); *The London Merchant's Mirror, or the Tradesman's Guide, being Tables for the ready casting up Bills of Exchange* (1705), (we may observe here that Harris makes use of numerous 'ready reckoners' and guides to methods of pricing throughout the *Lexicon Technicum*); *The British Hero*; *or a discourse shewing that it is the interest, as well as duty, of every Briton to avow his loyalty to King George on the present important crisis of affairs* (1715); *The Wickedness of the pretence of Treason and Rebellion for God's sake* (1715); *Astronomical Dialogues between a Gentleman and a Lady*: *wherein the Doctrine of the Sphere, uses of the Globes, and the Elements of Astronomy and Geography are explained* (1719), this book is one of many of the period which addressed the important subject of navigation: it was re-published in 1729 and 1766; and finally *The History of Kent* (1719).

From this brief survey of the life of John Harris we can see how he was intellectually well equipped to make an encyclopaedic survey of science, technology and the allied arts and crafts. The success with the London booksellers of his *Collection of Voyages and Travels* required editorial and compilation skills of a high order and disposed a group of ten publishers (see Title Page) to produce the *Lexicon Technicum* in the form of an initial folio volume which was followed-up by a further volume that served as a *supplement* to the first.

The Lexicon Technicum: *Volume I*

The first volume was published, as we have noted, in 1704. A small folio format was adopted (25cm x 31cm). The frontispiece depicts *Johannes Harris*, engraved by G White from the painting by R White: the author-compiler is portrayed in a cleric's gown and collar.[2]

Dedication

Harris dedicates the *Lexicon Technicum* to Prince George of Denmark (who did not become monarch until 1727) and thanks the Prince for accepting the patronage of his book. The Prince's courage is noted

(George I was the last British monarch to lead his troops in person into battle) and his other qualities are exalted: 'You exempt me from that hardship which attends most Dedicators, of Inventing the Virtues they celebrate.'

Subscribers
Eight closely-printed pages list over nine hundred subscribers: their names give some impression of the encyclopaedia-purchasing public in England early in the eighteenth century. Harris's subscribers include: Bishops, Gentlemen, Mathematicians, Merchants, Academics of Oxford and Cambridge, Booksellers, Lawyers, His Grace *Thomas* Lord Archbishop of Canterbury, Viscounts, Baronets, the Provost of Eton, and, most notably, Mr Isaac Newton, Master of the Mint.

The Preface to Volume I
A read through the Preface (see later) to the *Lexicon Technicum* tells us much about the task of compiling an encyclopaedia at the opening of the eighteenth century. A picture is given of an alert, well informed mind warming to the task of compilation. John Harris's great facility was in collecting and editing information. His appetite was omnivorous: no subject daunted him. He particularly relished abstruse propositions in mathematics in which, as we have noted, he was particularly skilled. In total, he collated over 8,200 entries in his first volume alone, some articles running to several pages. Harris appears to have worked largely unaided, save for the occasional assistance of an amanuensis. He is scrupulous in acknowledging his sources: 'I have usually taken particular care to give all Authors their just due, from whom I have taken any considerable part of my materials.'

Harris begins by setting out his purpose, referring to his *Lexicon* as a *dictionary*:

> That which I have aimed at, is to make it a Dictionary not only of bare Words but Things; and that the Reader may not only find here an Explication of the Technical Words, or the Terms of Art made use of in all the Liberal Sciences and such as border nearly upon them, but also those Arts themselves; and especially such Parts of them as are most Useful and Advantagious to Mankind.

In describing his work as a *dictionary* rather than an *encyclopaedia*, Harris was probably being mindful of the pioneering efforts that had gone before him. For example, exactly one hundred years previously Robert Cowdrey had published one of the very earliest dictionaries called a *Table Alphabeticall* (1604), whilst the honour of the first English *wordbook* to actually make use of the description *dictionary,* had gone to Henry Cockerman for his *English Dictionarie* (1623).[3]

That Harris certainly was aware of his predecessors is apparent from what follows: his assessment of contemporary dictionaries is however somewhat bleak and he cocks a snook in particular at the French:

> I found much less help from Dictionaries already published, [than] one would have expected from their Titles. *The Grand Dictionaire Des Arts & Sciences*, par M. de l'Academie Francoise, hath no Cuts [engravings] nor Figures at all, and is only a bare Explication of Terms of Art; and it seems rather to have been design'd to improve and propagate the *French Language*, than to inform and instruct the Human Mind in general.

Harris's main objection is that dictionaries of the period are too filled up with what he calls *simple terms* such as dog, cat, horse and sheep. He also protests at the listing of the names of animals, vegetables, metals and minerals and then passing these off as the *terms* of arts and sciences. Returning to the French encyclopaedists he adds:

> *Mr Furetière's Dictionary* may be Books very well done in their way, and are certainly very useful for those who would be perfectly acquainted with the *French* Tongue; yet I did not find much assistance from them, with regard to my Design.

Harris also distances himself from recent works by his fellow countrymen:

> And much less help was there to be had from a Book called, The *New World of Words*, or *an Useful English Dictionary*; for there I found the Compiler had the Ill Luck to Collect many *Foreign* Faults, and to understand little or nothing of the *Arts* and *Sciences* himself.

The use of illustrative material was rare in eighteenth-century encyclopaedias. Harris resolved to change this state of affairs. From the outset he conceived that line illustrations would form an integral part of his medium of communication. He gives the reader four whole-page plates which illustrate: *Air Pump, Anatomy, Architecture* and *Fortification*. Pride of place is reserved for two fold-out plates describing a *water pump* and a *Microscope*, true to the spirit of *technical Harris*. The feeling that the *Lexicon Technicum* is an illustrated book is however conveyed most by the many diagrams and tables which are numerous throughout the text. This innovative aspect of Harris's work was highly significant for the style and visual imagery of later encyclopaedias: it was taken up by Ephraim Chambers and flourished in the hands of Denis Diderot and his collaborators.

Harris closes his review of the work of others by stating that it is not his purpose to disparage their performance or to build himself a reputation on their ruins but 'to Acquaint the Reader with the plain truth of Things, without Favour or Affection'.

The many references in the Preface to science, mathematics, and technology fall outside the scope of our survey, but we can turn to *Fortification* to gain an insight into Harris's working method:

> As to *Fortification*, I have consulted the best Books and Drawings; and I believe the several Parts of a fortified Place are pretty well described, and the Plate belonging to this *Art*, at the End of the Book, comprehensive, and done according to the best of the Modern Ways.

Concerning the presentation of the plates in general in the *Lexicon Technicum*, Harris provides his readers with a section headed *Notes to the Plates*: which is placed at the end of Volume I and consists of *descriptions* and *explanations* of technical terms. We may note in passing that the integration of text and illustrations proved problematic to the early encyclopaedists, not least to Denis Diderot the Chief Editor of the *Encyclopédie* (who, by way of comparison, includes no less than seventeen plates on the subject of Fortification).

Introducing the plate on Architecture Harris, writing in an age when the unacknowledged use of sources was common, demonstrates the integrity with which he identified his sources:

> The Figures also of the Five Orders of Pillars in *Civil Architecture*, are I hope, as instructive as they cou'd be of that Size [all five orders are reproduced on a single page]; and as for the Terms of Art here, I took from the best Authors, as *Vitruvius*, *Vignola*, *Palladio*, and the *Parallel*, by Mr *Evelyn*.

In compiling his *Lexicon Technicum* Harris did more than consult books. He busied himself deriving information wherever he could at first hand. This is particularly apparent in the reference to *naval architecture*, our last quotation from the *Preface*: this text reveals Harris's resourcefulness in making direct, personal observations of the subjects he wished to record and describe in his text:

> And as to Ship, I have endeavour'd to be very full and particular in describing the several Parts of her, both in the *Dock*, when *Building*, and when *Rigg'd*, and *under Sail* at Sea: For I have consulted the best *Draughts*, *Sections* and *Models* I could get a sight of, have got what helps I could from Captains and Masters of Ships, and have often gone on Board my self, to get the more ready Knowledge of this Affair.

Harris's enthusiasm for making personal observations in order to improve his understanding of practical matters concerning technology and the mechanical arts, perfectly anticipates the spirit of enquiry that characterised Diderot's researches of such subjects for inclusion in the *Encyclopédie*: Diderot, and his collaborators, similarly inspected places of work and studied construction models to improve their understanding of design and technology.

Architecture in Volume I of the Lexicon Technicum

Volume I of the *Lexicon Technicum* runs to several hundred pages. These are unnumbered but the alphabetic sequence is adopted, itself something of an innovation in 1704. Articles on architectural subjects occur throughout Volumes I and II and we later conclude our survey with a listing of all the principal subject headings in this group.

A reading of the entry *Architecture* reveals how the polymath, scholar-cleric John Harris perceived this subject in the first decade of the eighteenth century and the aspects he chose to highlight and present to his readers.

Harris first defines architecture in terms of its different categories:

> Architecture, is the Art or Science of Building, and is usually divided into *Civil* and *Military*. *Civil Architecture* teaches how to make any kinds of Buildings, Palaces, Churches, or Private Houses: *Military Architecture* instructs us in the best way to fortify Cities, Camps, Sea-Ports, or any other Places of Strength.

The classical orders are then introduced:

> The Ancients at several Times, and on several Occasions, established 5 *Orders of Architecture*, that is, 5 sorts of Pillars, viz. the *Tuscan*, the *Dorick*, the *Ionick*, the *Corinthian*, and the *Composite* or *Roman Order*.

Each of these specific terms is further defined in the text under separate headings to which the reader is directed. This system of *cross-referencing* is maintained throughout the *Lexicon Technicum* with remarkable consistency. The article elaborates on the differences between the orders and explains the meaning of the terms: *Column, Base, Capital, Entablature, Architrave, Frise,* and *Cornice*. Two additional terms are introduced as evidence of further examples of 'Ancient Orders', namely, *Caryatides* and *Persick*. Turning to the definition of the latter, now archaic expression, we are given:

> Persick *Order of Architecture*, is where the Bodies of men serve instead of Columns to support the Entablature, or rather the Columns are in that form.

In a passage that owes much to the writing of Sir Henry Wotton[4], the ordering principles of architecture are next stated:

> The Rules of Architecture require, that in a well-built Fabrick there should be *Solidity*, *Convenience*, and *Beauty*; to which some writers add, *Order*, *Disposition*, *Proportion*, *Decorum*, and *Œconomy*: And these Eight make the necessary Parts of *Architecture*.

One of Harris's aims, as outlined in his Preface, is that the reader should 'find here an Explication of *Technical* Words, or the Terms of Art'. The article on *Architecture* affords a good illustration of the manner in which this particular aim is typically fulfilled. Each architectural term is defined in everyday words so as to progressively inform the reader of the art of creating architecture from built forms:

> *Solidity*, implies the choice of a good Foundation and good sound Materials to work with. *Convenience* consists in so ordering and disposing the Parts of an Edifice, that they may not hinder or embarass one another. *Beauty* is that agreeable Form and pleasing appearance which exhibits to the Eye of the Spectator. *Order* . . . gives each part of the Building a convenient Bigness, whether we consider them apart, or with Relation to the Whole. And *Disposition* they make the due Ranging and agreeable Union of all the Parts.

Harris turns to Vitruvius for a definition of *Proportion*:

> *Proportion* is the Relation that all the Work hath to its Parts . . . And to express the Relation that many things have to one another, as to their Greatness, and the different number of their Parts, Vitruvius indifferently uses those three Words; *Proportion*, *Eurythmy*, and *Symmetry*; the two last of which are of much the same sense with the first.

Decorum is identified with *Decency*: the definition given (see inventory of letterpress articles) evokes images of fine country residences, but Harris does not allow himself to be unduly swayed by romantic imagery and pays due regard to such practical matters as room planning and the admission of light:

> *Decorum* or *Decency* . . . consists in making the whole Aspect of the Fabrick so *correct*, that nothing shall appear but what is founded upon, and approved of by some Authority . . . *Decorum* teaches you to have a Regard to these 3 things; *Design*, *Custom*, and *Nature*. *Design* makes us choose . . . Dispositions and Proportions . . . *Custom*, makes us adorn the Porches and Entries into such Houses as are within Rich and Magnificent. And the Regard we have to the *Nature* of the

> Places, makes us pitch upon different Prospects for different parts of a Building; as for example, we expose Bed-chambers and Libraries to the Morning Sun: Winter Apartments to the *West*, and Closets of Pictures, etc. to the *North*, because they require an equal Light.

The final entry on *Architecture* focuses on financial matters which have all the relevance today they had when first written nearly three hundred years ago:

> *Œconomy* teaches the Architect . . . to have regard to the Expences that are to be made, and the Quality of the Materials, near the Places where he builds, and to take his Measures rightly for the Order and Disposition; *viz.* to give the Fabrick a convenient *Form* and *Magnitude*.

The Lexicon Technicum: Volume II (Supplement)

The Title Page for Volume II of the *Lexicon Technicum* is the same as that for Volume I but with *English Dictionary* emboldened and printed in red ink: Harris is given his courtesy title of Secretary to the Royal Society; the names of eight publishers are given, including five of the original ten; the work is dedicated, as noted earlier, to Lord Cowper; and the date of publication is given, MDCCX.[5]

At the start of his researches Harris planned a two-volume work which he makes clear in his Introduction:

> When I first began this work, I foresaw the Design could not be accomplish'd in One Volume; but however, I thought it more proper to Publish one Volume first, than to defer Printing *anything* at all of it till the Whole was finished: For as it is easie to see, that new Matter will continually occurr in a Design of this nature, and, consequently, that there can be no such thing as a *Perfect Book* of this Kind, so I thought it better to send out an Imperfect one than none, and to afford *some* Help to Mens Improvement in Philosophy and Mathematical Studies, rather than leave them to stay four or five Year for a more *Compleat* one.

Harris expresses his appreciation to 'the Learned World' for receiving his earlier endeavours so well: the success of the previous volume is reflected in the new list of subscribers now occupying twelve full pages. Collectively they make a more splendid group than the original list of subscribers, although the names of some subscribers occur in both lists. Harris's new list is liberally enriched with the names of Dukes, Earls, Viscounts and Baronets. The less exalted but more typical subscribers include: booksellers, merchants, surgeons, apothecaries, instrument makers and Jeremy Clark, organist of St. Paul's.

Building in Volume II of the Lexicon Technicum

The article Building from Volume II of the Lexicon Technicum is a good one with which to illustrate Harris's treatment of a technical subject. He begins with a sobering caution that suggests little has changed over the intervening years in the building trades concerning human nature: 'There are some good Rules and Directions about Building which . . . 'tis very proper every Gentleman shou'd be acquainted with, that he may not be cheated or imposed upon by his Workmen.' Harris then instructs the gentleman-builder in the art of good building construction, commencing, appropriately, with the foundations:

> First, with respect to the Foundation, care ought to be taken, that the Cellars are dug, if there be any, or if none, after the Trenches are dug in which the Walls are to stand, to try all the Foundations either by a Crow, a Rammer, or which is best, with a *Borer*, such as Well Diggers, Miners, etc. use, whether they are thoroughly sound, and are fit to bear the Weight that is to be laid upon them.

The article continues in a similarly informed manner on the bonding of brickwork with a rule-of-thumb for mortar mixes:

> The Walls also must be well wrought and *bonded* as they call it: In order to which the Mortar must be well made of good well burnt Lime and sharp Sand; the usual Proportion is . . . a Bushel of Lime to one and an half of Sand.

Brickwork construction is considered in greater detail in the next passage with the vagaries of weather in mind:

If the Bricks are laid in very hot dry Weather, and it be some small Piece of Work which you would have very strong, it will be worth while to dip every Brick as 'tis laid in a Pan of Water; and it will be of good use in great Work to throw Water on the Walls, after the Bricks are laid . . . And the Walls also should be covered in Summer Time, to keep them from drying to hastily, as well as in Winter Time from Rain, Snow, or Frost, which last is a great Enemy to all kinds of Mortar.

The closing section of the article *Building* offers guidance concerning the preservation of timber:

> When Timber is laid on Brick-work, as Torsels for Mantle Trees or Lintels for Windows, Templets for Girders, etc. let it be always laid in *Loam*, for that is a great Preserver of Timber, which Mortar will eat and corrode: The same thing should be done with the Ends of Joists and Girders, some Workmen dawb their End with Pitch, to preserve them from the Mortar.

Our survey of the *Lexicon Technicum* concludes with an inventory of all the letterpress articles on architectural subjects in Harris's two volumes.

Letterpress articles on architectural and allied subjects extracted from John Harris's *Lexicon Technicum*

VOLUME I

ABACUS
ACOUSTICKS
ACROTERIA
ALCOVE
AMPHIPROSTYLE
AMPHITHEATRE
ANAGLYPHICK ART
ANGLE
ANNULETS
ANTES
ANTESTATURE
ANTICKS
ANTIPAGMENTS
APOPHYGE
AQUEDUCT
ARÆOSTYLE
ARCHITECTONICK
ARCHITECTURE
ARCHITRAVE
AREOSTYLE
ASTRAGAL
ATTICK
AVENUE

BACULE
BALLESTER
BALLON
BALLUSTRADE
BANDELET
BANQUETTE
BARACK
BASE: FORTIFICATION, ARCHITECTURE
BASILICA
BASKETS
BASS-RELIEF
BASTION
BASTON
BATTERY
BATTLEMENTS
BERME
BEVEL
BLINDS
BONNET
BONNET À PRISTRE
BOYAU
BREACH
BREAK GROUND
BREAST-WORK
BRIDGE FLYING
BULWARK
BUST
BUTTRESS

CAPITAL OF A BASTION
CAPITAL
CAPONNIERE
CARIATIDES
CARNATION
CARYATIDES
CASCAN
CASEMATE
CASERN
CATACOUSTICKS
CATHETUS
CAVALIER
CAVIN
CAZEMATE
CHAMBRANLE
CHANDELIERS
CHANEL OF THE IONICK
CHAPITERS
CHEMIN DE RONDS
CHEMISE
CINCTURE
CIRCUMVALLATION
CITADEL
CLAYES
COFFER
COLOUR
COLUMN
COMPARTIMENT
COMPOSITE ORDER
COMPOSITION
CONCRETION
CONGE
CONSOLE
CONTRA MURE
CONTRAVALLATION
CORBEILS
CORDON
CORINTHIAN ORDER
CORNICHE
CORONA
CORRIDOR
CORTIN
COUNTER-APPROACHES
COUNTER-BATTERY
COUNTER-BREAST-WORK
COUNTER-FORTS
COUNTER-GUARDS
COUNTER-MINE
COUNTERSCARP
COUNTER SWAL-LOWS TAIL
COVERT-WAY
CROWN-WORKS
CROWN'D
CURTAIN
CYMATIUM

DADO
DARK TENT

DARKENED
DECORUM
DEFENCES
DEFILE
DEGRADATION
DEMI-BASTION
DEMI-GORGE
DENTILS
DESSEIN
DIACOUSTICKS
DIASTYLE
DIPTERON
DISTANCE OF THE BASTIONS
DISTEMPER
DOME
DONJON
DORICK ORDER
DOUCINE
DRAPERY
DRAUGHT COMPASSES
DRIP
DROPS
DURABLE FORTIFICATION
DYE
DYPTERON

ECHINUS
EMBRASURE
EMBRASURES
EMPATTEMENT
ENCEINTE
ENFILADE
ENFILE
ENTABLATURE
ENVELOPE
EPAULE
EPAULMENT
EPISTYLE
ESCALADE
ESPAULEMENT
ESPAUL
ESPLENADE
ETCHING
EURITHMY
EUSTYLE
EYE

FACE, ARCHITECTURE
FACE OF A BASTION
FACE OF A PLACE
FALSE-BRAYE
FASCIA
FASCINES
FASTIGIUM
FILET
FLANK
FLANKT ANGLE
FLAT CROWN
FOLIAGE
FORT
FORTIFICATION
FORTIN
FRAISES
FREEZE
FRESCO
FRIZE
FRONT: PERSPECTIVE, FORTIFICATION
FRONTON
FURRING
FUSAROLE

GABIONS
GALLERY
GLACIS
GORGE: ARCHITECTURE, FORTIFICATION
GOTHICK
GULBE
GUTTÆ
GUVE

HALF-MOON
HARMONY
HEPTAGON
HERSE
HERSILLON
HEXASTYLE
HOLLOW-TOWER
HORN-WORK
HURDLES
HYPERTHYRON
HYPOTRACHELION

ICHNOGRAPHY, FORTIFICATION
ICHNOGRAPHY, PERSPECTIVE
INTERCOLUMNIATION
INTERCOLUMNS
INTERIOR TALUS
INWARD FLANKING-ANGLE
IONICK ORDER
ISLES

JET D'EAU

LACUNAR
LARMIER
LINE
LIST
LISTEL
LIZIERE
LODGMENT
LUNETTES

MADRIER
MANNER
MANTELETS
MERLON
METOPA
MILITARY ARCHITECTURE
MINE
MOAT
MODEL
MODILLIONS
MODULE
MOINEAU
MONOTRYGLYPH

MOSAICK WORK
MOYENAU
MUSKET-BASKETS
MUTULE

NICHE
NORMAL
NUCLEUS

OBELISKE
OCTOSTYLE
OPEN FLANK
ORDER
ORDONANCE
ORILLON
ORLE
ORNAMENTS
ORTEIL
ORTHOGONIAL
ORTHOGRAPHY
OUT-WORKS
OVAL

PALLIER
PALLISADOES
PARAPET
PEDESTAL
PEDIEMENT
PERAMBULATOR
PERIPTERE
PERISTYLE
PERSICK ORDER
PERSPECTIVE:
AERIAL, LINEAL,
PRACTICAL,
SPECULATIVE
PICKET
PIED-DROIT
PIEDOUCHE
PIGMENTS
PILAE
PILLAR
PILLASTERS
PLACE OF ARMS
PLANCERE
PLANE,
FORTIFICATION
PLANE:
GEOMETRICAL,
HORIZONTAL
PLASM
PLAT-BAND
PLATFORM:
ARCHITECTURE,
FORTIFICATION
PLINTH
PLINTHUS
POLYGON EXTERIOR
PORTICO
POSTERN
PROFIL
PROJECTURE
PRONAOS
PROPERTY
PSEUDODIPTERON
PYCNOSTYLE

RAKED TABLE
RAMPART
RASANT LINE OF
DEFENCE
RAVELIN
REASON
REDENT
REDOUBT
RELAIS
RELEIFE
RETIRADE
RETRENCHMENT
REVERSED TALON
RIDEAU
ROMAN ORDER
RONDEL
ROOD

TO SAIGNER A MOAT
SAPPE
SAUCISSONS
SCAMILLI IMPARES
SCARP
SCENOGRAPHY
SCIOGRAPHY
SCOTIA
SCROWLES
SECTION OF A
BUILDING
SHOULDRING
SILLON
STAR
STEREOBATA
STEREOGRAPHY
STRUCTURE
STYLOBATA
STYLOBATON
SWALLOW-TAIL
SYMMETRY
SYSTYLE

TABLE: PROJECTING,
RAKED
TÆNIA
TAILLOIR
TALON
TALUS,
FORTIFICATION
TALUS: EXTERIOR,
INTERIOR
TECHNICAL
TEINT
TENAILLE
TENEMENT
TERRE-PLAIN
TETRASTYLE
THEODOLITE
TONDINO
TORE
TORUS
TRABEATION
TRAVERSE
TRENCHES
TRIGLYPH
TRINGLE
TROCHILE
TROPHY
TUSCAN ORDER
TUSCAN WORK
TYMPAN

TYMPAN OF AN ARCH

UNION

VESSELS
VISUAL POINT
VOLUTA

WAY OF THE ROUNDS
WINGS

ZOCCO
ZOCLE

VOLUME II

ABACUS
ABREUVOIRS
ADZ
ANCHORS
ANNULETS
ANTÆ
ANTIPÆGMENTS
ANTIQUE
APERTIONS
ARCHES
ARCHITECTURE
ASHLAR
ASPECT DOUBLE
ATTITUDES

BACK AND BOTTOM NAILS
BALDACHIN
BALLISTERS
BANK
BARBICAN
BASIL
BATTEN
BEAD
BEAM
BEARER
BEDMOULDING
BINDING JOYSTS
BOULTINE
BRACE
BRADS
BREST
BREST SUMMERS
BRICKS
BRIDGE OF COMMUNICATION
BUILDING
BUTMENTS

CAMBER BEAM
CANTALEVERS
CARCASE
CARTONS
CARTOUZES
CASEMENT
CASTING OF DRAPERY
CAUKING
CAULICOLI
CAVAZION
CAVETTO
CEMENT
CHAPEL
CHAPTRELS
CIMA
CIMATUM
CLAMP-NAILS
CLAMP'D
CLARO-OBSCURO
CLASP-NAILS
CLENCH-NAILS
CLOUT-NAILS
COLAPTICE
COLONNADE
COMMUNICATION
COMPARTITION
COMPARTMENT
COMPOSITE
COMPOSITION
CONTRASTE
CORBETT
COUCH
COVING-CORNISH
CRAYONS
CROWN-POST
CRYPTOGRAPHY
CULVERTAIL
CUPULO

DESIGN
DIAGLYPHICE
DILAPIDATION
DISTANCE OF THE EYE
DOG-NAILS
DORMANT TREE
DORMER
DOUBLE ASPECT
DOVE-TAILING
DRAGON BEAMS

EASEL-PIECES
EASEMENT
EAVESLATH
ENCAUSTICE
ENCOLAPTICE
EXTREMITIES

FACE PROLONG'D
FACIA'S
FLUTES
FLYERS
FLYING BRIDGES
FORNICATION
FRESCO
FURRING
FUST

GABLE-END OF A HOUSE
GAGE
GAIN
GIRDERS
GLYPHICE
GOTHICK
GOTHICK MANNER
GRAND GUSTO
GROTESQUE
GROUND PLATES
GROUPPE
GUERITE

HANCES
HATCHING
HEADS
HEIGHT OF THE EYE
HIP-ROOF
HIPS
HYPERTHRE

IMPOSTS
INLAND
INLANTAL
INSULATA COLUMNA
INTAGLIO'S

JOISTS
JOYSTS

KEYSTONE
KING PIECE

LAYMAN
LEAD-NAILS
LEDGERS
LINE OF THE SECTION
LOCAL COLOURS
LUTHERNS

MAGAZINE
MANNER
MANSE
MASSES
MEMBRETTO
MESSUAGE
METOPS
MITRE
MOAT
MODELL
MORESK-WORK
MORTISE
MOSAIC WORK
MUNIONS

NAVE
NEWEL

OGEE
OPENING OF THE TRENCHES
ORDONANCE
OVOLO

PAINTING
PALLIFICATION
PALLISADES
PARADIGRAMMATICE
PARASTÆ
PARGETING
PEDIMENT
PEERS
PIAZZA'S
PIEDROIT
PILÆ
PITCH
PLANE
PLASTICE
PORTRAITS
PRICK-POSTS
PRINCIPAL POSTS
PRINTS
PROFILE
PROSTYLE
PULVINATA
PUNCHINS
PURLINS
PUTLOGS

QUARTERS
QUIRK
QUOINS

RAFTERS
REFECTORY
REPOSE
RIGLET

SAPHETA
SCENOGRAPHICK
SCHEAME
SCRATCH-WORK
SCRIBING
SCULPTURE
SELL
SOFITTO
STILES
STILOBATUM
STRIÆ
STRIGES
STRUT
STUCCO

TAILLE DOUCE
TALUS SUPERIOR
TASSELS
TENAILLE
TENON
TESSELATA
THORUS
TIGE
TOMICE
TRANSOM
TRAVERSE
TRIMMERS

VIVO

WATER TABLE
WINDING-STAIRS

ZACCO
ZOPERUS

Postscript

The *Lexicon Technicum* contains several hundred letterpress articles on architectural subjects which admirably fulfil Harris's stated intention of 'explaining not only the terms of art, but the arts themselves'. Harris himself emerges as a punctilious editor-compiler. His material is well researched, carefully assembled, and is presented in a clear, objective style. The use of illustrations is innovative and generous, for the period, and served as an exemplar for later compilers. Numerous cross-references throughout the text supplement the major articles, and bibliographies further inform the reader. Harris is a dispassionate writer, free from personal eccentricity who spurns ridicule and satire. He is generally forward-looking, turning his back on mystics and alchemists. He embraces the new order epitomised by Newton's science and mathematics which he explains to the reader with the authority of a specialist. He was alert to advances in technology and recognised the importance of its application to the arts, crafts and manufactures.

The *Lexicon Technicum* deserves its high reputation. It is a remarkable one-man achievement. Volume I had to be reprinted in 1708 and at the time of Harris's death (1719) the work had progressed through three editions and a supplement. Two more editions (1724 and 1736) and a further supplement took the work down to 1744 by which time competition from Ephraim Chambers' *Cyclopaedia* brought the *Lexicon Technicum* to the end of its run, and in France the inception of the mighty *Encyclopédie* was about to be announced.

Scholarly success did not confer material benefits on John Harris who died a complete pauper. His companion and benefactor John Godfrey bore the expenses of the funeral arrangements and had the mortal remains of his good and worthy friend laid to rest at Norton Church, Kent.

5 *Lexicon Technicum* Volume I

Title Page

Dedication

The Preface

Lexicon Technicum:

O R, A N

Universal English Dictionary

O F

A R T S

A N D

S C I E N C E S:

Explaining not only

The *T E R M S* of *A R T*,

B U T T H E

A R T S Themselves.

By *JOHN HARRIS*, M.A. F.R.S.

L O N D O N:

Printed for *Dan. Brown*, *Tim. Goodwin*, *John Walthoe*, *Tho. Newborough*, *John Nicholson*, *Tho. Benskin*, *Benj. Tooke*, *Dan. Midwinter*, *Tho. Leigh*, and *Francis Coggan*, MDCCIV:

TO

His Royal Highness.

GREAT SIR,

I *Justly esteem it a Peculiar Happiness to me, that You were pleas'd to Accept of the Patronage of this Work: For hereby You exempt me from that hardship which attends most Dedicators, of Inventing the Vertues they Celebrate.*

But tho' Your Intrinsick Excellence and Real Worth takes away all Possibility of Flattery, yet being convinced that whatever I can say, falls short of your Merit: I Blush as much at my Incapacity of giving You Your just Due, as I should at ascribing more than he deserved, to another.

Your Early Years, GREAT SIR, *shew'd Your Skill and Conduct in Arms, as well as the Greatness of Your Courage and Love for Glory, and 'twas an uncommon Happiness Your Bravery was more than once blest with, by Your own Personal Valour to save Your Royal Brother, Rout the Enemies Troops, and change the Fate of the Battel.*

After this, when Your Happy Alliance with Your Royal Consort, *Our present Gracious and Most Excellent* QUEEN, *had made You* Ours, *tho' You were always ready to offer Your Self to Command Our Armies or our Fleets, yet when we deny'd our Selves the Happiness of Your Conquering for us Abroad, You set Your Self to shew us an Illustrious Example of Virtue and Goodness at* Home; *and prov'd Your Self to be as much above* Pride *and* Revenge *here, as You would have been without* Fear *or* Surprize there.

Great within Your Selves Alone, Your Royal consort

and Your Self, *like the best of the Old* Roman *Generals and Consuls, liv'd Retired, but neither unactively nor unusefully: Virtue and Piety You taught, by the best way of Recommending it, Your own Bright, Great and Glorious Examples: And raised up the Conjugal Happiness to such a degree of Perfection, as the World never knew before, and which could You have been* Unbeloved, *would have rendred You almost the Envy of every one.*

Then MIGHTY SIR, *did You employ Your Noble Mind in Studies of the greatest Use and Benefit to Mankind here,* Mathematicks *and* Mechanicks; *and as Your Closet was always the Resort of the usefully Learned and Ingenious, so were You still their Encourager and Patron.*

But when the Providence of God called Your Illustrious Princes to Empire, *and Your Self to Publick Business and Mighty Employments, You soon* Both *shew'd You cou'd Command with as much Wisdom and Conduct as You could before Obey with Resignation, and make Our Nation Truly Happy by the Gracious and Excellent Administration of Your Government.*

The Glorious Effects of which that we may long Enjoy, and that Our Gracious QUEEN, *and* YOUR ROYAL HIGHNESS *may be Blessed with Issue to Inherit those* Great Dominions, *that I doubt not but God hath taised up* HER SACRED MAJESTY, *to secure in lasting Peace and Happiness to us, is*

GREAT SIR,

The most Sincere and Constant Prayer

of Your ROYAL HIGHNESS'S

Most Obedient and Most Humble Servant,

John Harris.

THE

PREFACE.

THE best Account I can give of the following Work, will be to lay before you in a short View what it contains, wherein it differs from other Books which may seem to be of the same Nature, and from whence I have collected the Substance of it. That which I have aimed at, is to make it a Dictionary not only of bare *Words* but *Things*; and that the Reader may not only find here an Explication of the *Technical* Words, or the Terms of Art made use of in all the *Liberal Sciences*, and such as border nearly upon them, but also those *Arts themselves*; and especially *such*, and *such* Parts of them as are most Useful and Advantageous to Mankind. In this, which was the chief of my Design, I found much less help from Dictionaries already published, then one would have expected from their *Titles*: *Chauvin's Lexicon Rationale*, or *Thesaurus Philosophicus*, is a well Printed Book, and the Figures are finely Graved; but 'tis too much filled with the School Terms, to be usefully instructive; and is as defective in the Modern Improvements of Mathematical and Physical Learning, as it abounds with a Cant which was once mistaken for *Science*.

The *Grand Dictionäire Des Arts et Sciences, par M. de l'Academie Francoise*, hath no Cuts nor Figures at all, and is only a bare Explication of Terms of Art; and it seems rather to have been design'd to improve and propagate the *French Language*, than to inform and instruct the Humane Mind in general. And, which I have often wonder'd at, 'tis filled every where with *Simple Terms*, so that you are told what a Dog, a Cat, a Horse and a Sheep is; which, tho' it may be useful to some Persons who did not know *that* before, and may shew very well, that such Descriptions can be given in *French*; yet how bare Names of Animals and Vegetables, of Metals and Minerals, can be reckoned as *Terms of Art*, and consequently make the greatest part of a *Dictionary of Arts and Sciences*, I confess I cannot see. And therefore, tho' this and Mr. *Furetiere's Dictionary* may be Books very well done in their way, and are cartainly very useful for those who would be perfectly acquainted with the *French* Tongue; yet I did not find much assistance from them, with regard to my Design.

And much less help was there to be had from a Book called, The *New World of Words, or an Universal* English *Dictionary*; for there I found the Compiler had the *Ill Luck* to Collect many *Foreign* Faults, and to understand little or nothing of the *Arts* and *Sciences* himself.

Mr. *Ozanam's Dictionaire Mathematique*, is indeed as good a Book as *Vitalis's* is a bad one; and had *Mathematicks* been a Science I was the least acquainted with of any other, I should have been generally as well supplied from *Ozanam*, as deceived by *Vitalis*; whose last Edition in two Volumes at *Rome*, after so many Years time to consider upon the Matter, is I think, not better than the First, because it hath more Matter, and less to the Purpose.

The *Chymical* and *Physical Dictionaries* of *Johnson, Castellus* and *Blanchard*, have a great many Words and Terms that are not to be met with elsewhere: And the last hath had four Editions in our own Language; but tho' many things are well enough done in him, yet some can hardly be said to be so; so that in many Places I have been obliged to put his Name to what my *Amanuensis* or *Assistant* transcribed from him, lest the Reader shou'd mistake it for my own Words.

I Write not this only to disparage the Performances of others, or to build my self a Reputation on their Ruines: But I think my self obliged to Acquaint the Reader with the plain truth of Things, without Favour or Affection; that so he may be informed where to meet with satisfaction in his Inquiries and Reading, and where not; and perhaps if this were oftner done, both the Time and the Expence of gaining true Knowlege would be much shortned.

There are some other lesser Dictionaries which are of good use, and which have been serviceable to me on Occasion, which I shall mention below; but I must next acquaint the Reader, That *much the greater part* of what he will find here, is collected from *no Dictionaries*, but from the best Original Authors I could procure in all Arts and Sciences, and is the Result of some Years Labour and Consideration.

I have been very Full and Particular in the *Mathematicks*, because 'tis the only Solid Foundation on which a Useful Enquiry into Nature and all Physical Learning, can possibly be built; and because 'tis also of the greatest Use and Advantage to Mankind in all respects.

In *Geometry*, under the Name of each Figure, you have the essential Properties of it briefly and plainly demonstrated; and the Application of it to practice shewed: So that by the help of very easie References from one place to another, you will find the Demonstration of all such useful and important Proportions, in that Noble Science, as are usually given by *Geometrick* Writers.

Under such Words as *Parabola, Ellipsis, Hyperbola*; also in the *Conick Sections* you will find the Properties of each Figure or *Section* demonstated, with Methods for their Description on a Plain; as also the Properties of the *Cycloid, Conchoid, Logarithmic-Line, Ciffoid, Quadratrix* and *Spiral-Lines, etc.* Some general Considerations of the

Nature of *Asymptotes*, the Nature and Properties of *Catacaustick* and *Diacaustick* Figures, of the *Involute* and *Evolute* of the *Linea Celerrimi Descensus*, *etc*. And under the Word *Construction*, you have the Construction of *Cubick* and *Biquadratick Equations* by the *Parabola*; together with the Investigation of *Baker's Central Rule*, and its Use and Application.

Under such general Words as *Trigonometry*, *Surveying*, *Spherical-Geometry*, *Projection*, *etc*. you will find Entire Treatises on these Heads; and which, if I mistake not, are as short and plain, as any yet extant.

In *Algebra* you will meet under the proper Heads, every thing that is usually found in Treatises of this Nature, and perhaps something more. And under *Fluxions* you have a succinct Account of the *Nature* and *Algorithm* of them, and some Improvements which are not to be had elsewhere. And in several Places, under the proper Words, you have an Account of what we now call the *New Methods*, *or Universal ways of Investigation*; as particularly, a Method of drawing of *Tangents* to all sorts of *Curves*, a Method *de Maximis et Minimis*, of finding the *Centres of Gravity and Oscillation*; of finding the *Unciæ*, *etc*. All the Parts also of *Common Arithmetick* are explained here, and its Application to *Anatocism*, *Compound Interest* and *Annuities*; together with the Doctrine of Surds, the Method of Extraction of Roots by *Converging Series*, and the entire Doctrine of Proportion, all kinds of Progression, Fractions, Logarithms and Decimals: And I have also from the best Authors I cou'd get, collected an Account of the Ancient Weights and Measures of all Nations, and adjusted them with our own; and have given you very large and useful Tables of the Values of all Modern or present Foreign Weights, Measures and Coins compared with our own. I have inserted also from Dr. *Wallis*, a large Account of *Sexagesimal Fractions*. You will find here also the Descriptions and Uses of both the Celestial and Terrestial Globe, the Doctrine of the Sphere, according to the New and Old Hypothesis; together with the Demostration and Practice of Spherick Projections *in Plano*; and most of the useful things in *Astronomy*; as the *Phœnomena*, *Paralaxes*, *Magnitudes*, *Motions* and *Distances* of the *Planets* and their *Satellites*; and in particular, the Incomparable Mr. *Isaac Newton's Theory of the Moon*; and a very large Account of *Comets* from the same Author: You will have also the Ground and Practice of *Dialling*, the Nature and Use of *Opticks*; the taking all *Heights* and *Distances*; the *Mensuration* of *Surfaces*; *Gauging*; the *Art* of *Chronology*, *Geography*, *Cosmography*, *Musick*, *etc*.

In *Gunnery* you have the Method of Shooting in Great Guns and Mortars, with Captain *Halley's* and *Anderson's* Tables; and as to *Fortification*, I have consulted the best Books and Drawings; and I believe the several Parts of a fortified Place are pretty well described, and the Plate belonging to this *Art*, at the End of the Book, comprehensive, and done according to the best of the Modern Ways.

The Figures also of the Five Orders of Pillars in *Civil Architecture*, are

I hope, as instuctive as they cou'd be of that Size; and as for the Terms of Art here, I took them from the best Authors, as *Vitruvius*, *Vignola*, *Palladio*, and the *Parallel* by Mr. *Evelyn*.

Navigation also is largely treated of, and the whole *Art* taught under the Words *Plain* and *Mercator's Sailing*, and *Traverses*: And as to the *Variation of the Compass*, I have given you a very full and exact Account of it from that Excellent Mathematician Capt. *Halley*, together with his Ingenious Hypothesis, for the Solution of its Variation; and Practical Rules to find the Variation of the Compass at Sea.

And as to a Ship, I have endeavoured to be very full and particular in describing the several Parts of her, both in the *Dock*, when *Building*, and when *Rigg'd*, and *under Sail* at Sea: For I have consulted the best *Draughts*, *Sections* and *Models* I could get a sight of, have got what helps I could from Captains and Masters of Ships, and have often gone on Board my self, to get the more ready Knowledge of this Affair; and I have compared it all with what we have already Printed of this Nature in Books and Descriptions of Ships; such as *Manwaring's Sea Dictionary*, *Boteler's Sea Dialogues*, *Philip's* Section of a First Rate, *etc*.

I have also from several Eminent Hands, and several ways, the *Laws of Motion*: As from Dr. *Wallis*, Mr. *Isaac Newton*, Monsieur *Varignon*, Mr. *John Keil*, *etc*.

And the Doctrine of *Mechanicks*, *Nature* and *Properties of Staticks*, and the *Pendulums*, and of *Projectiles*.

And I have been very large in that most Useful Science *Hydrostaticks*; giving an Account of the Principles and Practice of that *Art*, both Mathematically and Experimentally: And under the Word *Specifick Gravity*, I have given you all that the Honourable Mr. *Boyle* hath advanced on that Subject, in his *Medicina Hydrostatica*; to which a large Table of the *Specifick Gravities* of different Bodies is added; and all his *Hydrostatical Paradoxes* are inserted in their proper place.

I have described also, chiefly from Mr. *Durham*, the way of *Calculation* of *Automata*, or Clock and Watch-work; and explained the Terms of *Art* used in *Painting* and *Sculpture*.

As to *Physicks* and *Natural Philosophy*, and those admirable Helps to the Understanding of Nature, which *Geometry* applied to *Physical Enquiries*, hath of late afforded us, and to which indeed we are chiefly indebted to that Prodigious Mathematician Mr. *Isaac Newton*. I have endeavoured to give you every where the Marrow and Substance of it under proper Heads: And under the Term of Art or Word expressing any particular *Quality*, I have collected all I could meet with to explain it, and to clear up its Nature and Properties; as you will find at large under such Words as *Electricity*, *Solidity*, *Elasticity*, *Effluviums*, *Magnetism*, *Light* and *Colours*, *etc*. As to which last I'm sorry I had time to take no more from Mr. *Newton's* Excellent Book of *Opticks*, lately published; but what I could, I have inserted, as I had before done the Substance of what he had published in the *Philosophical Transactions*, on that most Noble

Subject; and which convinces us that we were all mistaken in our Notions about it before, for want of proceeding in a right Method of Enquiry.

The *Phœnomena* of the *Rain-bow* or *Iris*, are here accounted for from the Learned and Ingenious Capt. *Halley*, now *Savilian Professor* of Geometry in *Oxon*.

The Account of *Snow* I give you from Dr. *Grew*: and one of *Ice* from the *French*.

I have Collected what I could meet with as to *Sound*; but I wish that Quality were a little better considered.

You have here a very full Account of all the *Phœnomena* and *Properites* of the *Air* and *Atmosphere*, as its *Gravity*, *Spring* or *Elasticity*, *etc*. and a full Description of, and the Use of such Instruments as have been invented to enable us to judge of them; as the *Barometer*, *Thermometer*, *Hygrometer*, *etc*. all which is chiefly from the Honourable Mr. *Boyle* and the *Philosophical Transactions*.

The Account of *Springs* and *Fountains*, is from Capt. *Halley*, and Dr. *Woodward*: And from the former of these Learned and Ingenious Gentlemen, I have given you a new Theory of *Tides*, or a Solution of the *Phœnomena* of the Ebbing and Flowing of the *Sea*, which is very plain, certain and intelligible, and built on Mr. *Newton's* Principles.

In *Botany* I have been as large as I could be, without giving Description of Plants, which is contrary to my Design: But you have here from our Mr. *Ray*, *Morison*, and Monsieur *Tournefort*, I believe, a pretty exact *Botanick Lexicon*, which was what we *really* wanted before: Together with an Account of all the several *Kinds*, and *Subalternate Species* of *Plants*, and their *Specifick Differences*; in which I have followed Mr. *Ray's* Method, as appearing to me to the best and most Natural.

A Table of *Fossils* I have given you from the Accurate Dr. *Woodward*, Professor of Medicine in *Gresham* College: And a Scheme of *Metals* and *Stones* from Bishop *Wilkins's* Real Character.

I have also given you a large Account of *Vegetation*, which is very curiously and exactly done from the Experiments and Observation of Dr. *Woodward*.

You have here also a good Account of the Nature and Property of the *Wind*, from Capt. *Halley*: And a Description of Mr. *Papins's Wind-Gun*, from Mr. *Boyle*.

From which last Excellent Gentleman, I have also taken what I say about the Nature and Properties of *Cold*.

And likewise from Capt. *Halley* and Mr. *Newton*, is collected all that which you will find under the Word *Heat*.

In *Chymistry*, the Knowledge of which is one great help towards the understanding of Nature, I have been large and particular; explaining the Chymical Principles, Vessels and Degrees of Fire, and have omitted no Process nor Operation of Use, that I could either meet with in Books, or get from my Friends; as the Reader will soon see by consulting the Book it self, under such Words as *Phosphorus*, *Bolonian Stone*, *Sympathetick Inks*, *Transmutation*, *etc.*

In *Anatomy* I have been very large and full, describing all Parts of an Humane Body, both internal and external: And tho' indeed the Figures for this part are much less than I could have wished them to be; (for the Book hath so much out-run the Expence the Undertakers at first proposed, that they would not be brought to have larger Cuts;) yet I hope the Descriptions in most Places, will prove tolerably Accurate and Instructive; especially under such general Words, as *Blood*, *Circulation*, *Bones*, *Heart*, *Ear*, *Eye*, *Vision*, *Arteries*, *Veins*, *Bile*, *Lympha*, *Chylification*, *etc.* in all which, I have consulted the Best Authors.

In *Logick*, *Metaphysicks*, *Ethicks*, *Grammar*, *Rhetorick*, *etc.* I have been designedly short; giving usually the bare meaning only of the Words and Terms of Art, with one or two Instances to explain them, and illustrate them.

In *History* and *Chronology*, you have what properly belongs to them as *Arts*; as an Account of the *Civil* Computation of Time; the Original and Reduction, one to another, of several *Æra's*, *Epocha's*, *Periods*, *etc.*

In *Heraldry* I have given you the entire Art of *Blazening* and *Marshalling* a *Coat of Arms*; and explained all the *Ordinaries*, *Charges*, *Bearings*, *etc.* by *Figures*: But have said nothing of Families (any further than that such a Coat belongs to such a Name) my Design being only to explain the *Art* and its *Terms*.

As to the Description of the *Mathematical* and *Philosophical Instruments* some of them (as the most useful) are largely done, but others as briefly: For it would have taken up a large Volume to have described them all; so for many of them, the Reader would not have been much the better. But the *Globes* and *Quadrants* are largely treated of, as are *Telescopes*, *Microscopes*, *Baroscopes*, *Hygroscopes*, and the *Pneumatick-Engines*, or *Air-Pumps*; because these are of vast Use and Benefit to Mankind, and have served to improve and raise up the Knowledge of Nature to that good height it is now arrived to, and I hope will carry it yet much further.

And as I have usually taken particular care to give all Authors their just due, from whom I have taken any considerable part of my Materials, without Partiality, so I have designedly done Justice to such Ingenious and Industrious Artificers, as do truly deserve the greatest Encouragement for their Skill and Accuracy in the making of those Instruments: Such are Mr. *John Rowley*, *Mathematical Instrument-maker*, under St. *Dunstan's-Church* in *Fleet-street*: Mr. *Yarwell*, and Mr. *Marshal*, *Perspective-Makers* in *Ludgate-street*: Mr. *Hawkesbee*, who makes *Air-Pumps* and all *Pneumatick-Engines*, in *Giltspur-street* near *West-Smithfield*: And Mr. *John Patrick*, the *Torricellian Operator* in the *Old-Bailey*, who makes all Kinds of *Barometers* and *Thermometers*; as you will find I have done in the proper Places: And I can't here omit mentioning the Ingenious Mr. *Wilson*, which I could not do in the Book, because those Sheets about the *Microscope* were Printed off before I had seen Mr. *Wilson* or his Glasses: But I must now do him Justice to say, That of all the *Microscopes* I have seen for Commodiousness, various Uses, Portability and Cheapness; I never met with anything like Mr. *Wilson's* Glasses. They are particularly described in the *Philosphical Transactions*, *N.*

As to the *Law Part* of this Dictionary, I did my self actually consult the best Books and Dictionaries I could get recommended to me; and from thence I transcribed abridgedly all that seem'd necessary to be inserted here; and since that I have had it carefully examined and corrected by a Gentleman of known Ability in that Profession.

And thus having given you a short Account of what you may expect to find in this Work, and which may perhaps satisfie the Reader that it is a Book useful to be *read carefully over*, as well as to be consulted like other Dictionaries occasionally: I shall next fairly acquaint you wherein it is defective, and what Improvements may be made of it hereafter, in an additional Volume.

The Catalogue, Right-Ascensions, Declinations, *etc.* of the *Fixed-Stars*, is very imperfect; the Reason of which is, That Mr. *Flamstede*, upon whom I thought I might depend, was pleased to refuse me any Communication of that kind; else I would have given those things under the Name of each Star or Constellation.

There are also, I doubt, here and there some Words which my *Amanuensis* or *Assistant* transcribed from other Dictionaries, which are not so well explained as they should be, and which (among so many Thousand Words as I had to range into Order) have escaped a Review; but I have corrected as many of them as I could in each Sheet as the Book was Printing off.

I would have had also at the End of the Book, a Particular *Alphabet* for each *Art* and *Science* by itself; and some more and larger Copper-Plates in *Anatomy*, and of the *Outside*, *Rigging*, and the *Section* of a *Ship*. But the Undertakers could not afford it at the Price proposed, the Book having swelled so very much beyond the Expectation: But whatever Alterations, Amendments, Improvements and Additions shall be hereafter, as I question not but many of the Latter sort Time will produce, if God please to Bless me with Health and Leisure, there shall all be Printed in a Volume by themselves, and so by no means be prejudicial to the first Impression.

6 *Lexicon Technicum* Volume I

Architectural subjects portrayed in the letterpress articles

Our inventory of letterpress articles in the series *The Encyclopaedic-Dictionary in the Eighteenth Century: Architecture, Arts and Crafts* is composed of subjects derived from various branches of architecture and allied arts and crafts. The following short list of headwords is provided here to give the reader an outline impression of the range of material that is included in the letterpress articles:

Architectural terms, arch construction, building construction, building materials, cabinet work, carpentry, classical orders, decorative art, design principles including aesthetics, domes, drainage, fortification, furnishings, furniture, gardening, joinery, landscape terms, military art, mouldings, ornament, painting, perspective, sanitation, sculpture, surveying, terminology of the building trades and crafts, tools and equipment used in building construction, and vaults.

Lexicon Technicum Magnum:

Or, an UNIVERSAL

English Dictionary

OF

Arts and Sciences.

VOL. I.

A

ABACUS, in Architecture, is the four square Table that makes the Capital on the Top of a Column, especially those of the *Corinthian* Order, and is a Drip or *Corona* to the Capital. It supports the neither Face of the Architrave and whole Trabeation. In the *Corinthian* and *Composite* Orders, the Corners of it are called, the *Horns*; the middle part the *Sweep*, and the *Curvature*, the *Arch*; which commonly has a *Rose* carved in the middle.

ACOUSTICKS, are Medicines or Instruments which help the Hearing; as to the latter of which, the use of *Otoccustick* Trumpets is now very common; and no question Instruments of this kind are capable of Improvement. Dr. *Hook* in his Preface to his Micrography saith, 'tis not impossible to hear a Whisper at a fur'ong or ⅛th part of a Miles distance: That he knows a way by which 'tis easie to hear any one speak thro' a Wall of a Yard thick: And that by the help of a distended Wire the Sound may be propagated to a very considerable distance almost in an Instant.

ACROTERIA or *Acroters*, signifie in Architecture those sharp Pinacles and spiry Battlements, which stand in Ranges about flat

Buildings, with Rails and Balasters: They signifie also a sort of *Pedestal* to support Statues.

ALCOVE, in Arichtecture is a part of a Chamber separated by an *Estrade* or Partition made by Columns and other corresponding Ornaments, in which is placed a Bed of State; or sometimes Seats to entertain Company.

AMPHIPROSTYLE, a Term in Architecture for a kind of Temple of the Ancients, which had four Columns in the Front, and as many in the Face behind.

AMPHITHEATRE, is a Place built for Acting and seeing of Publick Spectacles and Stage-Plays: The Common Theatre was but Semicircular, but this was built entirely round, or in form of a perfect Circle.

ANAGLYPHICK ART, is the Art of Carving, Engraving, or Embossing.

ANGLE *of or at the Center*, in Fortification, is that which is made by the Concourse of two Lines drawn from the Angles of the Figure.

ANGLE *of the Circumference*, in *Fortification*, is the mix'd Angle made by the Arch which is drawn from one Gorge to another.

ANGLE *of the Counterscarp*, is made by the two Sides of the Counterscarp meeting before the middle of the Curtain.

ANGLE *of the Curtain*, or *Angle of the Flank*, is that which is made by or contained between the Curtain and the Flank in any Piece of Fortification.

ANGLE DIMINISHED, in Fortification, is that which is made by the meeting of the outermost sides of the *Polygon*, and the *Face* of the *Bastion*.

ANGLE *of the Exterior Figure*, or *The Angle of the Polygon*, is that which is formed at the Point of the *Bastion*, by the meeting of the two outermost Sides or Bases of the *Polygon*.

ANGLE *of the Interior Figure*, in Fortification, is that which is made in the Center of the Bastion by the meeting of the innermost sides of the Figure.

ANGLE FLANKING, is that which is made by the meeting of the two *Rasant* Lines of Defence, *viz*. the two Faces of the Bastion prolong'd.

ANGLE *flanking inward*, is the Angle made by the Flanking-Line and the Curtain.

ANGLE FLANK'D, by some called, the *Angle of the Bastion*, in Fortification, is that which is made by the two *Faces*; being the utmost part of the Bastion most exposed to the Enemies Batteries, and therefore by some is called *the Point of the Bastion*.

ANGLE *forming the Flank*, is that which consists of one *Flank* and one *Demi-Gorge*.

ANGLE *forming the Face*, in Fortification, is that which is composed by one *Flank* and one *Face*.

ANGLE *of the Moat*, in Fortification, is that which is made before the *Curtain*, where it is intersected.

ANGLE RE-ENTERING, or *Re-entrant Angle*, in Fortification, is that which retires inward toward the Place.

ANGLE SAILANT, is that which advanceth its Point towards the *Field*. This is called *Sortant* or *Viff*.

ANGLE *of the Shoulder*, in Fortification, is that which is constituted by the Lines of the *Face* and *Flank*.

ANGLE *of the Tenaille*, or the outward *Flanking Angle*, called also *Angle Mort* or the *Dead Angle*, *Angle Rentrant* or *Angle Inwards*; is made by the two Lines *Fischant*, in the Faces of the two Bastions extended till they meet in an Angle towards the Curtain; and is that which always carries its Point in, towards the Work.

ANGLE *of the Triangle*, in Fortification, is half of the *Angle of the Polygon*.

ANNULETS, in Architecture, are small square Parts turned about in the *Corinthian Capital*, under the *Echinus* or Quarter-round.

ANTES, in Architecture, are Square Pilasters which the Ancients used to place at the Corners of the Walls of their Temples.

ANTESTATURE, is a small Retrenchment made of *Pallisadoes*, or Sacks of Earth set up in haste to dispute with the Enemy the remainder of a piece of Ground, part whereof hath been already gained.

ANTICKS, in Architecture, are certain carved Works representing divers odd shapes of Men, Beasts, Birds, Fishes, Flowers, *etc.* which being rudely determin'd and form'd one out of another, serve as an Ornament to that whole Fabrick, and afford a grateful variety to the Eye of the Beholder.

ANTIPAGMENTS, in Architecture, are the Garnishings of Posts or Pillars.

APOPHYGE, in Architecture, signifies that part of a Column, where it seems to fly out of its Base like the Process of a Bone in a Man's Leg, and begins to shoot upward. But this *Apophyge* is really no more than the Rings or Ferrils heretofore fasten'd at the Extremities of wooden Pillars to keep them from splitting, and afterward imitated in Stone work.

AQUEDUCT, is the Term in Architecture for a Conduit, or Work to convey Water (without force of Engines) to any Place.

ARÆOSTYLE, in Architecture, is a sort of Edifice where the Pillars are set at a great distance one from another.

ARCHITECTONICK, is that which builds a thing up regularly according to the Nature and Properties of it. Thus that Plastick Nature, Power, Spirit, or whatever it be called, which hatches the *Ova* of Females into Living Creatures of the same species, is by some called the *Architectonick Spirit.*

ARCHITECTURE, is the Art or Science of Building, and is usually divided into *Civil* and *Military. Civil Architecture* teaches how to make any kinds of Buildings, Palaces, Churches, or Private Houses: *Military Architecture* instructs us in the best way to fortify Cities, Camps, Sea-Ports, or any other Places of Strength.

The Ancients at several Times, and on several Occasions, established 5 *Orders of Architecture*, that is, 5 sorts of Pillars, *viz.* the *Tuscan*, the *Dorick*, the *Ionick*, the *Corinthian*, and the *Composite* or *Roman Order*. (*Which see under those Words.*) The Difference between which *Orders*, consists in the *Column*, with its *Base* and *Capital* and the *Entablature*, that is, the *Architrave*, *Frise* and *Cornice*; for these are the Parts which constitute the *Order*, and each one hath its proper and peculiar Measures.

Besides these 5 Ancient Orders, some Authors mention two more, as that of the *Caryatides* and the *Persick.* (Which see.)

The Rules of *Architecture* require, that in a well-built Fabrick there should be *Solidity*, *Convenience*, and *Beauty*; to which some Writers add, *Order*, *Disposition*, *Proportion*, *Decorum*, and *Oeconomy*: And these Eight they make the necessary Parts of *Architecture*. *Solidity*, implies the choice of a good Foundation and good sound Materials to work with. *Convenience* consists in so ordering and disposing the Parts of an Edifice, that they may not hinder or embarass one another. *Beauty* is that agreeable Form and pleasing appearance which it exhibits to the Eye of the Spectator. *Order* (they say) gives each part of the Building a convenient Bigness, whether we consider them apart, or with Relation to the Whole. And *Disposition* they make the due Ranging and agreeable Union of all the Parts.

Proportion is the Relation that all the Work hath to its Parts; and which every one separately hath to the *Idea* of the whole: For amongst Works that are perfect, from any particular part, we make a certain Judgement of the greatness of the whole Work: *v. gr.* the *Diameter of a Pillar*, or the length of a *Triglyph*, gives us a right Idea of the whole to which they belong. And to express the Relation that many things have to one another, as to their Greatness, and the different number of their Parts, *Vitruvius* indifferently uses those three Words; *Proportion*, *Eurythmy*, and *Symmetry*; the two last of which are of much the same Sense with the first.

Decorum or *Decency* comes next to be considered, which consists in making the whole Aspect of the Fabrick so *correct*, that nothing shall appear but what is founded upon, and approved of by some Authority: And they say, *Decorum* teaches you to have a Regard to these 3 things; *Design*, *Custom* and *Nature*. The Regard to *Design* makes us choose (for Instance) other Dispositions and Proportions for a Palace, than a Church. The Respect we Pay to *Custom*, makes us adorn the Porches and Entries into such Houses as are within Rich and Magnificent. And the Regard we have to the *Nature* of the Places, makes us pitch upon different Prospects for different parts of a Building; as for example, we expose Bed-chambers and Libraries to the Morning Sun: Winter Apartments to the *West*, and Closets of Pictures, *etc.* to the *North*, because they require an equal Light.

Oeconomy teaches the Architect (they say) to have a regard to the Expences that are to be made, and the Quality of the Materials, near the Places where he builds, and to take his Measures rightly for the Order and Disposition; *viz.* to give the Fabrick a convenient *Form* and *Magnitude*.

ARCHITRAVE, is the chief Beam in any Building, and the first Member of that which is called the *Entablature*; *viz.* that part of a Stone Pillar which is above the *Capital*, and below the *Frize*. And 'tis called the *Reason-piece* or *Master-Beam* in Timber Buildings; but in Chimneys it is called the *Mantle-piece*, and over the Jambs of Doors or Lintels of Windows *Hyperthyron*. The *Greeks* called it the *Epistyle*.

AREOSTYLE, in Architecture, is a Building where the Columns stand a little too thick, as some say; or as the *French Dictionary of Arts and Sciences* expresses it, *Loin à loin*, that is at a convenient distance.

ASTRAGAL, in Architecture, is by the *French* called the *Talon*, by the *Italians*, *Tondino*, and is a kind of half *Tore*, sometimes wrought (in the Richer Orders) like a kind of overcast Hemm or Edge to the larger *Tore*; which frequently is placed between, as in the *Ionick Base*, with two *Scotia's* or *Trochiles*; and sometimes, but rarely, just about the Plinth of the Base. Sometimes it is taken for the Collar or Cincture next the *Hypotrachelium* and Diminution of a Pillar listed on both Edges; and it runs also round under the *Echinus* of the *Ionick*. We find it also sometimes dividing the *Fascia* of the Corinthian Architrave, where it is wrought in Chaplets, and Beads or Berries; and sometimes 'tis used both above and below the *Lists*, and joining immediately to the Square or Die of a Pedestal, where the *Stylobata* is introduced. *Freart's Parallel of Architecture.*

ATTICK, a kind of *Order* in Architecture, after the manner of the City of *Athens*. *Vitruvius* makes it the Name of a *Basis*, which the Modern Architects have since given to the *Dorick Pillar*. In our Buildings, the *Attick* is a little *Order* placed upon another much greater, having, instead of Pillars, nothing but *Pilasters* of a particular form or fashion.

AVENUE (in Fortification) is an opening or inlet into any Fort, Bastion, or such like Place.

B

BACULE (in Fortification) is a kind of Port-Cullis or Gate made like a Pit-fall with a Counter-poise, and supported by two great Stakes. 'Tis usually made before the *Corps de Gards* advanced near the Gates.

BALLESTER or *Balluster*, in the Capital of the *Ionick Column*, is the lateral part of the Scroll, which makes the *Curl-tuft*.

BALLON, in Architecture, is taken for a round Globe, or top of a Pillar.

BALLUSTRADE, in Architecture, signifies a Row of little turn'd Pillars, high enough for a Man to rest his Elbows on, fixed upon a Terrass, or upon the top of a Building; or to make any Separation between one part of it and another.

BANDELET, is a *French* Term for one of the Ornaments in Architecture, which they call also *Regle*: 'Tis greater than the *List*, but a little less than what they call a *Platte Bande*, it encompasses the Pillar quite round about like a Ring.

BANQUETTE, in Fortification, is a little Foot-pace or Elevation of Earth in form of a Step, or the bottom of the *Parapet*, or that which the Musqueteers get up to discover the *Counterscarp*, or to fire upon the Enemies in the *Moat* or in the *Covert-way*. These *Banquets* are generally a Foot and a half high, and almost three Foot broad.

BARACK or *Baraque*, is an Hutt like a little Cottage, for Soldiers to lie in, in the Camp: Formerly those for the Horse were called *Baracks*, and those for the Foot *Hutts*; but now *Barack* is used for both indifferently. They are made usually when the Soldiers have no Tents, or when an Army lies long in a Place in bad Weather, because they keep out Cold and Wet much better than Tents. They are generally made by fixing four strong forked Poles in the Ground, and then laying four others across them; then they build up the Walls with Sods, Wattles, or what the Place affords: The Top is either Planked or Thatched, or covered with Turf, as they have Convenience.

BASE, in Fortification is the exterior side of the Polygon, *viz.* the Imaginary Line which is drawn from the *Flank'd Angle* of a *Bastion* to that which is opposite thereto.

BASE or *Basis*, in Architecture, is the Foot of a Pillar that sustaineth it, or that part which is under the Body, and lies upon the Pedestal or Zocle when there is any: Therefore 'tis not used for the lowest part of a Column, but for all the several Ornaments or Mouldings which reach from *Apophyges* or the rising of the Shafts of Pillars to the *Plinth.*

BASILICA, in the Ancient Architecture, was a great Hall which had two Ranges of Pillars, and two Isles and Wings with Galleries over them. These great Halls were at first made for the Palaces of Princes, afterwards they turned into Courts of Justice, and at last into Churches.

BASKETS *of Earth*: See *Corbeills.*

BASS-RELIEF, See RELIEFE.

BASTION, in Fortification, is now what was anciently called a *Bulwark*: It consists of two *Faces*, as many *Flanks*, formerly a *Gorge*. It is usually made at the Angles of Forts, of a large heap of Earth, sometimes lined with Stone, but usually faced with Sods. The Lines terminating it are *two Faces*, *two Flanks*, and *two Demigorges*. The Union of the two Faces makes the outmost Angle, called the *Angle of the Bastion*. The Union of the two Faces to the two Flanks makes the Side Angles, called the *Shoulders* or *Epaules*; and the Union of the two other Ends of the Flanks to the two Curtains forms the Angles of the Flanks.

BASTION *Solid*, are those that have their Earth equal to the height of the *Rampart* without any void Space toward the Centre.

BASTIONS *Void* or *Hollow*, are those that have a *Rampart* and *Parapet* ranging only round about their Flanks and Faces, so that a void Space is left toward the Centre, and the Ground is there so low, that if the *Rampart* be taken no Retrenchment can be made in the Centre, but what will lie under the Fire of the Besieged.

FLAT BASTION, is that which is built on a Right Line.

CUT BASTION, is that which makes a Re-entring Angle at the Point; and is sometimes called,

BASTION *with a Tenaille*, whose Point is cut off and makes an Angle inwards, and two Points outwards: This is done when Water, *etc.* hinders carrying the Bastion to its full extent, or when it would be too sharp.

BASTION *Composed*, is when the two Sides of the Interior Polygon are very unequal, which makes the *Gorges* also unequal.

BASTION *Deformed*, is that which wants one of its *Demi-Gorges*, because one side of the Interior Polygon is so very short.

BASTION *Demi*, hath but one Face and Flank, and is usually before a Horn-work or Crown-work: This is also called an *Epaulment*.

BASTION *Double*, is that which on the Plain of the great Bastion hath another Bastion built higher, leaving 12 or 18 Feet between the Parapet of the lower and the Foot of the higher.

BASTION *Flat*; if the Distance between the Angles of the Interior Polygon be double the usual length, then a Bastion is made in the middle before the Curtain; but it generally hath this disadvantage, that unless there be an extraordinary Breadth allowed to the Moat, the returning Angle of the Counterscarp runs back too far into the Ditch, and hinders the Sight and Defence of the 2 opposite Flanks.

BASTION *Regular*, is that which hath its due proportion of Faces, Flanks and Gorges.

BASTON, a *French* word in Architecture, the same with the *Tore*.

BATTERY, in Fortification, is a place raised on purpose where Cannon are planted, from thence to play upon the Enemy; the Platform on which they are fix'd being made of Planks that support the Wheels of Carriages, so as to hinder the Weight of the Cannon from sinking them into the Ground, and incline somewhat towards the *Parapet* to check the Recoiling of the Pieces.

BATTERIES *of a Camp*, are Places to plant great Guns on, and are usually surrounded with a *Trench* and *Pallisadoes* at the Bottom, as also with a *Parapet* on the Top, having as many Holes as there are *Pieces of Artillery*; and with two *Redouts* on the Wings, or certain Places of Arms capable of covering the Troops that are appointed for their Defence.

In all *Batteries* the open Spaces left to put the Muzzles of the Great Guns out are called *Embrazures*, and the Distances between the *Embrazures Merlons*. The Guns are generally about 12 Foot distant one from another, that the Parapet may be strong, and the Gunners have room to work.

BATTERY *Sunk or Buried*, is when its Platform is sunk or let down into the Ground so that there must be Trenches cut in the Earth against the Muzzles of the Guns, for them to fire out at, and to serve for *Embrasures*. This sort of Battery which the *French* call *En Terre*, and *Ruinante*, is generally used on the first making of Approaches, to beat down the Parapet of any Place.

BATTERIES *Cross*, are two Batteries which play athwart one another upon the same thing, forming there an Angle, and beating with more Violence and Destruction, because what one Bullet shakes the other beats down.

BATTERY *d'Enfilade*, is one which scours or sweeps the whole length of a Strait Line.

BATTERY *en Echarp*, is that which plays obliquely.

BATTERY *de Revers*, or *Murdering Battery*, is one that bears upon the Back of any Place.

BATTERY *Joint*, or *par Camerade*, is when several Guns play at the same time upon one Place.

BATTLEMENTS, are the Tops of the Walls of Buildings, made in the Form of *Embrasures* and *Merlons*, as in Fortified Places.

BERME, in Fortification, is a little Space of Ground three, four, or five Foot wide, left without, between the Foot of the *Rampart* and the side of the *Moat*, to receive the Earth that rolls down from thence, and to prevent its falling into the *Moat*. Sometimes for more Security the *Berme* is *Pallisadoed*.

BEVEL, an Instrument well known to Builders, Carpenters, and Bricklayers, and used for adjusting of Angles.

BLINDS, in Fortification, are certain Pieces of Wood, or Branches of Trees laid across from one side of the Trench to the other, to sustain the *Bavins* or Hurdles laden with Earth; and serve to cover the Pioneers from above, and are commonly used when the Works are carried on towards the *Glacis*, and when the Trench is extended in Front toward the Place.

BONNET, in Fortification, is a certain Work raised beyond the *Counterscarp*, having two Faces which form a Saliant Angle, and as it were a small *Ravelin* without any Trench. The Height of this Fortification is three Foot, and it is environ'd with a double Row of Pallisadoes, ten or twelve Paces distant from each other. It hath a Parapet three Foot high; and is like a little advanc'd *Corps du Guard*.

BONNET *à Pristre*, or *the Priest's Cap*, in Fortification, is an Out-work having at the Head three Saliant Angles, and two Inwards: It differs from the double Tenaille only in this, that its Sides instead of being Parallel are made like a Swallow's Tail, that is, narrowing or drawing close at the Gorge, and opening at the Head.

BOYAU, or *Branch of the Trenches*, in Fortification, is a particular Ditch separated from the main Trench, which in winding about encloseth different Spaces of Ground, and runs parallel with the Works and Fences of the Body of the Place; so that when two Attacks are made near one to another, the *Boyau* sometimes makes a Communication between the *Trenches*, and serves as a Line of Contravallation not only to hinder the Sallies of the Besieged, but also to secure the Miners. But when 'tis a particular Cut that runs from the Trenches to cover some Spot of Ground, it is then drawn parallel to the Works of the Place, that it may not be *Enfiladed*; that is, that the Shot from the Town may not scour along it.

BREACH, in Fortification, is the Ruins which are made in any part of the Works of a Town, *etc.* either by playing Cannon, or springing Mines, in order to storm the Place, or take it by Assault. They say, Make good the Breach, Fortifie the Breach with *Chevaux de Frise*, Make a Lodgment on the Breach; Clear the Breach, or move away the Rubbish of it, *etc.*

BREAK *Ground*, in Fortification, signifies, to begin the Works for carrying on the Siege about a Town or Fort.

BREAST-WORK, the same with *Parapet.*

BRIDGE *Flying*, a Term in Fortification: See *Pont Volant*.

BULWARK, much the same (anciently) with a *Bastion* in Fortification: Which see.

BUST, is a Term in Sculpture, signifying a *Figure* or *Statue* of but one half of a Humane Body; the Head, Shoulders and Breast appear, but no Arms, and it is made tapering from the Breast downwards.

BUTTRESS, is an Arch or Mass of Stone serving to support the sides of a Building, Wall, *etc.* on the outside: They are chiefly used in such Buildings as are of the *Gothick* Manner.

C

CAPITAL *of a Bastion*, in Fortification, is a Line drawn from the Angle of the Polygon to the Point of the Bastion, or from the Point of the Bastion to the middle of the Gorge. These *Capitals* are from 35 to 40 Fathom long, that is to say, from the Point of the Bastion to the Place where the two Demigorges meet.

CAPITAL, or *Chapitel*, or *Chapiter*, signifies in Architecture, the Top of a Pillar; and this is different, according to the different Orders.

CAPONNIERE, in Fortification, is a covered Lodgment of about 4 or 5 Foot broad, encompassed with a little Parapet of above two Foot high, which serves to support the diverse Planks laden with Earth.

This Lodgment is large enough to contain 15 or 20 Soldiers, and is usually placed upon the Extremity of the Counterscarp, having sometimes several little Embrasures made therein, commonly called *Mardresses*. They are generally on the Glacis, or in dry Moats.

CARIATIDES: See *Caryatides*.

CARNATION, is a Term in Painting, signifying such Parts of an Human Body as are drawn naked, without Drapery; or what express the bare Flesh; and when this is done Natural, Bold and Strong, and is well coloured, they say of the Painter, that his *Carnation* is very good.

CARYATIDES; an Order of Pillars in Architecture in the form of the Bodies of Women with their Arms cut off, and clothed in a Garment down to their Feet: These Figures do support the *Entablature*: There are some Columns of this Order in the *famous Building* at Bourdeaux, called the *Tutelles*, and in the great Hall of the *Swiss* Guards at the *Louvre* in *Paris*; where four *Caryatides* support a Gallery enriched with Ornaments very well cut.

The rise of this Order of Pillars was this; The Inhabitants of *Carya* in *Peloponnesus* did once treacherously join with the *Persians* against their own Country; but the *Grecians* got the Battle, and put all the Male Inhabitants of *Carya* to the Sword, reducing the City to Ashes. The Women they took Captives, and carried in Triumph: And the more to perpetuate the Memory of this Base Action, they would not permit these Women ever to put off their Triumphal Vestments; but ordered the Architects of those times to express them in that dress, as supporting the heavy weight of their Edifices, instead of usual *Columns*.

The *Persick Order* of Pillars had much the same Original: *Which see. Vitruvius lib.* 2. *c.* 1.

CASCAN, in Fortification, is a certain Hole or Hollow-place in form of a Well, from whence a Gallery dug in like manner under Ground is convey'd, to give Air to the Enemies Mine. Some of these *Cascans* are more hollow than others, being usually made in the Retrenchment of the Plat-form near the Wall.

CASEMATE, in Fortification, sometimes is the Well with its several Subterraneous Branches or Passages dug in the Passage of the Bastion till the Miner is heard at work, and Air given to the Mine.

Also a certain Vault made of Mason's Work, in that part of the Flank of a Bastion which is next the Curtain, on purpose to fire upon the Enemy, and to defend the Face of the opposite Bastion of the Moat.

Sometimes it consists of 3 Plat-forms, one above another, the *Terreplan* of the Bastion being the highest. Behind the Parapet, which fronts along the Line of the Flank, there are Guns placed laden with Cartridges of Small-shot to scour along the Ditch; and these are covered from the Enemies Batteries, by Earth-works, faced or lined with Wall, and are called *Orillons* or *Epaulments*. This is the best Defence a Place can have.

CASERN, in Fortification, is a little Room, or Lodgment, or Building, erected between the Rampart and the Houses of Fortified Towns; to serve as Appartments or Lodgings for the Soldiers of the Garrison: There are usually 2 Beds in each *Casern* for six soldiers to lie 3 and 3; but the 3d part being always on the Guard, there are but 4 left in the *Casern*, 2 in a Bed.

CATACOUSTICKS, or *Cataphonicks*, is the Science of Refelected Sounds, or which teaches the Doctrine and Properites of *Echoes*.

CATHETUS, in Architecture, is taken for a Line supposed directly to traverse the middle of a Cylindrical Body, as of a *Ballister* or Column. In the *Ionick* Chapiter it is also a Line falling perpendicularly, and passing through the Center or Eye of the *Voluta*.

CAVALIER, in Fortification, is a Heap or Mass of Earth raised in a Fortress, to lodge the Cannon for scouring the Field, or opposing a Commanding Work. These *Cavaliers* are sometimes of a Round, sometimes of a Square Figure, the Top being bordered with a Parapet to cover the Cannon therein mounted. There must be 12 Foot between Cannon and Cannon; and if they are raised on the Enclosure of any place, whether in the middle of the Curtin, or in the Gorge or Bastion, are generally 15 or 18 Foot high above the Terreplain of the Rampart.

A *Cavalier* is sometimes called a *Double Bastion*, and is designed to overlook the Enemies Batteries, and to scour their Trenches.

CAVIN, in Fortification, is a hollow place proper to favour the Approaches to a Fortress, so that one may advance therein under covert towards the Enemies, as it were in a Trench. If it be within Musket-shot, 'tis a place of Arms ready made to hand; and a convenience for opening the Trenches out of fear of the Enemies Shot.

CAZEMATE, a Term in Fortification: See *Casemate*.

CHAMBRANLE, an Ornament in Masonry and Joyners Work, bordering the three Sides of Doors, Windows and Chimneys. It is different according to the several Orders, and consists of three Parts, *viz.* the Top, called the *Traverse*, and the two Sides the *Ascendants*.

CHANDELIERS, in Fortification, are Wooden Parapets made of two upright Stakes six Foot high, which support divers Planks laid across one another, or Bavins filled with Earth. They are made use of in Approaches, Galleries, and Mines, to cover the Workmen, and to hinder the Besieged from forcing them to quit their Labours. The only difference between *Chandeliers* and *Blinds*, is, that the former serve to cover the Pioneers before, and the latter to cover them over Head.

CHANEL, in the *Ionick Capital*, is a part which is somewhat hollow under the *Abacus* after the *Listel*, and lies upon the *Echinus*, having its Contours or Turnings on each side to make the *Voluta's*.

CHAPITERS, in Architecture, is the Crown or upper part of a Pillar: Those that are destitute of Ornaments are called *Chapiters with Mouldings*; such as the *Tuscan* and *Dorick*; the first whereof is the most simple, having its *Abacus* Square without any Mouldings; but the *Abacus* of the other is crowned with an *Astragal* and three *Annulets* under the *Echinus*. All those that have Leaves and Carved Ornaments are term'd *Chapiters* with Sculptures, and the finest of them is the *Corinthian*, which is adorned with two Rows of Leaves, as also eight greater, and as many lesser *Voluta's* placed under a Body called a *Tympanum*. These are called usually *Capitals*.

CHEMIN *de Ronds*, in Fortification, is the way of the Rounds, or a Space between the Rampart and the low Parapet under it for the Rounds to go about. The same with the *Fausse Bray*.

CHEMISE, in Fortification, is a Wall with which a Bastion, or any other Bulwark of Earth is lined for its greater Support and Strength; or it is the Solidity of the Wall from the *Talus* to the Stone row.

CINCTURE, in Architecture, is that part which makes the middle of a Pillar; 'tis *Congè* in *French*; the Greeks called it *Apophyge*, because that part of the Pillar taking as it were a Rise, seems to fly from the Basis: It is no more than the Rings or Ferrils formerly used to preserve Wooden Pillars from splitting, and which they afterwards imitated in Stone Work.

CIRCUMVALLATION, or *the Line of Circumvallation*, in Fortification, is a Trench bordered with a Parapet, dug round about the Besiegers Camp within Cannon-shot of the Place, to hinder the Relief of the Besieged, and to stop Deserters: It is usually flank'd at the distance of a Musket-shot with *Redoubts* and other small Works, or with Field Forts raised upon the most eminent Posts: Care must be taken never to draw a *Line of Circumvallation* at the bottom of a rising Ground, lest the Enemy having seized on the Station should plant Cannon there, and by that means command the Line. This Line is usually about seven Foot deep and twelve broad.

CITADEL, is a Fort of 4, 5, or 6 Bastions, built somewhere near a City, that it may command it in case of a Rebellion; so that the City is not fortify'd on that part against the *Citadel*, tho' the *Citadel* is against the City. The most usual Form for *Citadels* is that of a Pentagon, a Square being too weak, and an Hexagon too big.

CLAYES or *Wattles*, a Term in Fortification. See *Hurdles*.

COFFER, in Fortification, is a hollow Lodgment athwart a dry Moat from six to seven Foot deep, and from sixteen to eighteen broad, the upper part being made of Pieces of Timber raised two Foot above the Level of that moat; which little Elevation hath Hurdles laden with Earth for its covering, and serves as a Parapet with Embrasures. The Besieged generally make use of these *Coffers* to repulse the Besiegers when they endeavour to pass the Ditch, and they are distinguished only by their Length from the *Caponnieres* which are likewise somewhat less in Breadth: And it differs from the *Traverse* and *Gallery*, in that these are made by the *Besiegers*, and the *Coffer* by the *Besieged*.

To save themselves from the Fire of these *Coffers*, the Besiegers *Epaule*, or throw up the Earth on that side towards the *Coffer*.

COLOUR, may be considered two ways, 1. As it is a Quality residing in the Body that is said to be so and so coloured, or which doth modify the Light after such and such a manner: Or 2. As more properly, the Light it self which so modified strikes upon the Organ of Sight, and produces that Sensation which we call *Colour*.

The Peripateticks assert *Colours* to be real Qualities, and inherent in the coloured Bodies, and suppose that Light doth only discover them, but not any effect of their Production.

Plato thought *Colour* to be a kind of Flame consisting of most minute Particles, very congruous to the Pores of the Eye, and darted against it from the Object. And some Moderns will have *Colour* to be a kind of Internal Light of the more lucid parts of the Object, darkened, and consequently altered by the various mixtures of the less Luminous Parts.

Others, as did some of the Ancient Atomists, maintain *Colour* not to be a Lucid Stream, but a Corporeal Effluvium issuing out of the coloured Body.

Others account for all *Colours* out of the various mixture of Light and Darkness; and the Chymists sometimes will have it arise from the Sulphur and sometimes from the Salt that is in Bodies; and some also from the third Hypostatick Principle, *Mercury*.

The *Cartesians*, who make the Sensation of Light to be the Impulse made on the Eye by certain solid but very minute Globules easily permeating the Pores of the Air and Diaphanous Bodies. These, I say, derive *Colour* from the various Proportion of the direct progress or Motion of these Globules, to their Circumrotation or Motion round their own Centres; by which means they are qualified to strike the Optick Nerve after distinct and divers manners, and so do produce the Perception of various *Colours*.

Dr. *Hook* in his *Micrographia* says, The Phantasm of *Colour* is caused by the Sensation of the oblique or uneven *Pulse* of Light; and that this is capable of no more Varieties than two, which arise from the two sides of the Oblique Pulse; so that there is in reality but two simple *Colours*, Yellow and Blue; from the Mixture of which, and a due proportion of

Black and White (that is, Darkness and Light) all *Colours* may be produced.

The incomparable Mr. *Isaac Newton* found by two Experiments on Prisms, that there is a great difformity in the Rays of Light; and that hereby the *Origin of Colours* may be unfolded.

An **Abstract** *of Mr.* **Newton's** *Doctrine concerning* **Light** *and* **Colours.**

1. Light consists of an infinite number of Rays right-lined and parallel, but of different degrees of Refrangibility when meeting with a different *Medium*.

2. Each Ray according to its degree of Refrangibility, when so refracted, appears to the Eye of a different Colour.

3. The least Refrangible Rays appear of a deep Scarlet, the most refrangible appear of a Violet Blue, the intermediate proceeding from Scarlet to Yellowish, then to light Green, and so to Blue.

4. The Colours arising from the different Refrangibility of Light are not only the more noted Colours of Red, Yellow, Green, Blue, but also all the intermediate degrees of Red to Yellow, of Yellow to Green, *etc.* differing as the degrees of sound from Grave to Acute; in which there are not only the Notes of common denomination, but also indefinite intermediate degrees of Sound which are as distinct different Sounds as the other.

5. Whiteness (such as the Sun's Light appears) containing all these degrees of Refrangibility, is consequently made up of all the abovementioned Colours.

6. Simple or Homogeneal Colours are such as are produced by Homogeneal Light or Rays, that have the same degree of Refrangibility; and mixt Colours are such as are produced by Rays of different Refrangibility.

7. Rays of the same Refrangibility produce the same Colour, which Colour is not alterable by repeated Refractions, only made more strong or faint as the Rays are united or scattered.

8. All Bodies appear of this or that Colour according as the Surfaces are adapted to reflect only the Rays of such a Colour or (at least) in more plenty than the rest.

A CATALOGUE of the **Simple Colours.**

1. *Spanish White*, made of *Chalk* and *Alumn* burnt together.

2. *Lapis Armenius*, supposed to be the same with the common *Blue Bice*.

3. *Ultramarine*, made of the bluest *Lapis Lazuli*, which is freest from Gold Veins, by Calcination.

4. *Smalt*, made of *Zaffer* and *Pot-ashes* calcin'd together in a Glass Furnace.

5. *Litmose*, supposed to be the Juice of a Plant.

6. *Indico*, by some said to be a kind of Mud adhering to the Froth about Reeds in *India*. Others says it is a Plant like *Rosemary* called *Inil*, growing in *Cambaia*, which is gathered and dry'd, then wetted with fair Water, and beaten to a Mud; this Operation being repeated, it is dry'd and sifted for use.

7. *Indian Ink*, whose Composition is supposed to be burnt *Rice*.

8. *Ceruss* or *Psymithium*, is the Rust of Lead made by a Vaporous Calcination.

9. *Masticot*, is a kind of improper *Calx* of *Tin*.

10. *Gambodia* or *Gutta Gambæ*, is supposed to be the inspissated Juice of a *Plant*, that comes from the *Indies*.

11. *Oker*, a sort of Natural Earth.

12. *Orpiment*, a fat inflammable Mineral, justly ranked amongst *Poysons* for its extream Corrosive Quality.

13. *Umber*, a native Earth.

14. *Red-Lead*, made of Litharge or burnt Lead by a Reverberatory Calcination, or of *Ceruss* put on a Platter over the Fire, which must be continually stirred till it has acquired a *Red-Lead* Colour.

15. *Burnt Oker*, is the common Yellow *Oker* burnt in open Fire.

16. *Cinnabar* or *Vermillion*, there are two sorts; Native, or the *Minium* of the Ancients, which is a Mineral that yields *Quick-silver*; whereof, and of *Sulphur* it chiefly consists; it is found in the Mines of *Istria*.

The *Factitious Cinnabar*, is that which we now use; and is made by a Sublimation of *Mercury* and *Sulphur*.

17. *Carmin*, made of *Cochineal*.

18. *Lake*, is made of *Flocks* dyed, or shavings of *Scarlet-Cloth*, or of the *Cochineel Insect*, or else of *Kermes-Berries*, their Tincture being extracted with a Lee of *Pot-ashes*, and then precipitated with a Solution of *Roch-Alum*. After the same manner a *Lake* may be made of a Plant or Flower: There is also another sort of *Lake* made of *Gum-Lac*, by extracting its Tincture with *Urine*.

19. *Sanguis Draconis*, is the Gum of a Tree which looks like dry'd Blood; 'tis brought out of several places in the *East-Indies*.

20. *Earth Reddle* or *Ruddle*, found in many places of *England*.

21. *Lamp-black*, made of *Soot* of *Rosin* or *Pitch* burnt in places built on purpose for it, that keep in the Smoak.

COLUMN, in Architecture, taken in the largest Sense, is a sort of Pillar of a round form, which serves to support or adorn a Building, consisting of a Base, a Shaft, or Shank, (which is properly the *Column*) and a *Chapiter*. *Columns* are different according to the several Orders of Architecture.

The *Tuscan* being the shortest and most simple, is seven Models long, comprehending its Base and Capital, and diminished a fourth part of its Diameter.

The *Dorick* ought to be seven and a half, or eight Diameters long; and its Base and Chapiter are somewhat more beautify'd with Mouldings.

The *Ionick Column* is nine Diameters long, and hath its Chapiters set off with *Voluta's*, or curl'd Scrolls, differing in that respect from others as well as in its Base, which is peculiar thereto.

The *Corinthian* is the Richest of all, having two rows of Leaves for the Ornament of its Chapiter, with Stalks or Stems, from whence shoot forth small Voluta's. Its length is ten Diameters.

The *Composit Column* is likewise ten Diameters long, and its Chapiter is made like that of the *Corinthian*, with the Angular *Voluta's* of the *Ionick*.

COLUMN, in Architecture, is properly that round long Cylinder or part of a Pillar which is called the *Shaft*, *Trunk*, *Fust*, the *Scapus Vivo*, *Tige*, *etc*. containing the Body of it from the Spire to the Base, or from the Astragal of the Base to the Capital.

COMPARTIMENT, is a regular orderly disposition of agreeable Figures all round any Picture, Map, Draught, *etc*. for its better Ornament.

COMPOSITE *Order*, in Architecture, which is also called *Italick* and *Roman*, because it was the Invention of the Ancient *Romans*, is so called, because it is *composed* of the other four *Orders*, viz. *Tuscan*, *Dorick*, *Ionick*, and *Corinthian*; the first is *composed* only of the *Ionick* and *Corinthian*.

COMPOSITION, in Painting, is used in the same sense with *Invention* or *Desein*: See *Desein*.

CONCRETION, is the uniting together of several small Particles of a Natural Body into sensible Masses, or *Concrete*, whereby it becomes so and so figured and determined, and is indued with such and such Properties.

CONGE, a Term in Architecture. See *Apophyge*.

CONSOLE, is a kind of Bracket or Shouldering-Piece in Building which hath a Projecture, and serves to support a Cornice, or to bear up Figures, Butts, Vessels, and other Ornaments of the like nature.

CONTRA MURE, in Fortification, is a little Wall built before another partition Wall to strengthen it, so that it may receive no Damage from the adjacent Buildings.

CONTRAVALLATION, or *The Line of Contravallation* in Fortification is a Trench guarded with a Parapet, and usually cut round about a Place by the Besiegers to secure themselves on that side, and to stop the Sallies of the Garrison; 'tis without Musket-Shot of the Town. So that the Army forming a Seige lies between the Lines of *Circumvallation* and *Contravallation.*

CORBEILS, in Fortification, are little Baskets about a Foot and a half high, eight Inches broad at the Bottom, twelve at the Top; which being filled with Earth, are frequently set one against another upon the Parapet, or elsewhere, leaving certain Port-holes, from whence to fire upon the Enemy under covert, without being seen by them.

CORDON, in Fortification, is a Row of Stones made round on the outside, and set between the Wall and the Fortress which lies in *Talu,* or a Sloap, and the Parapet which stands Perpendicular after such a manner, that this difference may not be offensive to the Sight; so that these *Cordons* serve only as Ornaments ranging round about the place; neither are they used but in Fortifications made with *Mason's* Work, the void Space being filled up with pointed Stakes in those that are raised with Earth.

CORINTHIAN *Order of Architecture*, is so called from *Corinth*, the place of its Invention. It observes the same Measures with the *Ionick*, and the difference between them is chiefly in the *Capital*: See *Order* and *Column*.

Vitruvius makes the height of the Capital to be two Modules, and this he divides into 7 parts, of which the first is for the *Abacus*, and the other six downwards is for the Ornament.

CORNICHE, or as we pronounce it *Cornish*, in Architecture, is the third and highest part of the *Entablature*, and is commonly used to signifie the uppermost Ornament of any Wainscot, *etc.* In reference to the *Pillar* 'tis different according to the different Orders of Architecture. In the *Tuscan* 'tis without Ornaments, and this Pillar of all others hath the least Mouldings. The *Dorick* is adorned with *Dentils* like the *Ionick*, and which sometimes hath its Mouldings cut into it. The *Corinthian* Pillar, hath of all others, the most Mouldings, and those very often cut with *Modillons*, and also sometimes *Dentils*. The *Composite* hath its *Dentils* and its Mouldings cut, with its Channels or Chamferings under its *Platfond*.

CORONA, in Architecture, is properly the flat and most advanced part of the Cornice, called by the *French Larmier*, and by us *the Drip*, because it defends the rest of the Work from Wind and Weather; but it is often taken by *Vitruvius* for the whole *Cornice*.

Corona, or the *Flat-Crown*, is also a particular Member in the *Dorick* Gate, made by so extraordinary an enlargement of the *Larmier* or *Drip*, that it hath six times more Breadth than Projecture. This sort of *Corona* is no where found amongst the Ancients, except only in the Writings of *Vitruvius*.

CORRIDOR, in Fortification, is the *Covert-way* lying round about the whole Compass of the Fortifications of a Place, between the outside of the Moat and the Pallisado's.

CORTIN, in *Fortification*, signifies the Wall or Distance between the Flanks of two Bastions.

COUNTER-APPROACHES, in Fortification, are Works made by the Besieged to hinder the Approach of the Enemy, and when they design to attack them in form.

COUNTER-BATTERY, is one raised to play against another.

COUNTER-BREAST-WORK, a Term in Fortification, the same with *False Bray*.

COUNTER-FORTS, are certain Pillars, and parts of the Walls of a Place, distant from 15 to 20 Foot one from another, which are advanced as much as is possible in the Ground, and join'd to the height of the *Cordon* by Vaults to support the Way of the Rounds, and part of the Rampart, as also to fortifie the Wall, and strengthen the Ground; nevertheless they are not at present much used except in large Fortifications.

COUNTER-GUARDS or *Envelopes*, in Fortification, are large heaps of Earth in form of a Parapet raised above the Moat, before the Faces and the Point of the Bastion to preserve them; and then they consist of two Faces, making an Angle Saliant and parallel with the Faces of the Bastion. If they are designed to cover one of the Faces of the Bastion, they are shaped like a Demi-Bastion, with a Parapet on the Face and Capital, but none on the Flank, which must lie open and be exposed to the Fire of the place: But there are few places now fortified after this manner by reason of the excessive Charge it requires.

COUNTER-MINE, in Fortification, is a Sub-terraneous Passage made by the Besieged in search of the Enemies Mine, to give Air thereto, to take away the Powder, or by any other means to frustrate the effect of it, though it should even happen to be fired by the Assailants.

COUNTERSCARP, in Fortification, is that side of the Ditch which is next the Camp, or properly the *Talus* that supports the Earth of the *Covert-way*; although by this Term is often understood, the *whole Covert-way* with its *Parapet* and *Glacis*. And so it is to be understood, when 'tis said, *The Enemy lodged themselves on the Counterscarp*.

COUNTER *Swallows Tail*, is an Out-work in Fortification, in the form of a single Tenaile, wider next the Place, *i. e.* at the Gorge than at the Head, or next the Campaign. The Sides of this *Counter Swallows Tail* are not so well flank'd from the Place, as those of the *true Swallows Tail*, and therefore 'tis not so good.

COVERT-WAY, in Fortification, is a Space of Ground level with the Field on the Edge of the Ditch, three or four Fathom broad, ranging quite round the Half-Moons, and other Works toward the Country. It is otherwise called *Corridor*, and hath a Parapet raised on a Level, together with its *Banquets* and *Glacis*, which from the height of the Parapet ought to follow the Parapet of the Place till it is insensibly lost in the Field. It hath also a Foot Bank. One of the greatest difficulties in a Siege, is to make a Lodgment on the Covert-way, because usually the Besieged Pallisade it along the middle, and undermine it on all sides: This is sometimes called, and that commonly, the *Counterscarp*, because 'tis on the Edge of it.

CROWN-WORKS, in Fortifications, are certain Bulwarks advanced toward the Field to gain some Hill or rising Ground, being composed of a spacious *Gorge*, and two Wings that fall on the *Counterscarp* near the Faces of the *Bastion*, so that they are defended by them, and present on the side of the Field, an entire *Bastion* between the two *Demi-Bastions*, whose Faces look towards one another. Those Works also have their *Half-Moons*, and are raised only to take up or secure some large space of Ground, or to defend the head of a Camp when it is intrenched.

CROWN'D *Horn-work*, is a *Horn-work* with a *Crown-work* before it.

CURTAIN, in Fortification, is the Front of the Wall of a Fortify'd Place between two *Bastions*; or the longest straight Line that runs about the Rampart, drawn from one Flank to the other, and bordered with a Parapet five Foot high, behind which the Soldiers stand to fire upon the Covert-way and into the Moat. Besiegers seldom carry on their Attacks against the *Curtain*, because 'tis the best flank'd of any Part.

CYMATIUM, a Member of Architecture; whereof one half is Convex, and the other Concave: There are two sorts of *Cymatiums*, *viz.* one called *Doucires* or *Right Gula*, the most advanced part of which is

Concave; and the other *Talon* or *Reversed Gula*, which hath its most advanced part Convex, but is hollow below, as the first is above.

D

DADO, a Term in *Architecture* by some Writers used for the *Dye*, which is the Part in the Middle of the Pedestal of a Column, between its Base and Cornice. 'Tis of a Cubick form, whence the Name *Dye* is given to it.

DARK *Tent*, is the Term which some Writers give to a Box made almost like a Desk with Optick Glasses, to take the Prospect of any Building, Fortification, Landskip, *etc.* This is a Portable *Camera Obscura*, or *Darkened Room*.

DARKENED *Room*, *Obscura camera*. See *Dark Tent*.

DECORUM or *Decency*, in Architecture, is a suiting all the Parts and Ornaments of a Building so as that they shall become the Station; as *Vitruvius* shews, when he speaks of adjusting the several Orders to their Natural Affections; and he, as Mr. *Evelyn* observes, would never have placed a *Corinthian Column* at the Entrance into a Prison, nor one of the *Tuscan* Order before the *Portico* of a Church.

DEFENCES, in Fortification, are all sorts of Works that cover and defend the opposite Posts, as *Flanks*, *Parapets*, *Casemates*, *Fausse Brays*: 'Tis almost impossible to fix the Miner to the Face of a Bastion till the Defences of the opposite one are ruined; *i. e.* till the Parapet of its Flank is beaten down, and the Cannon in all parts that can fire upon that Face which is attacked, are dismounted.

DEFILE, in Fortification, is a straight narrow Lane or Passage through which a Company of Horse or Foot can pass only in File, by making a small Front, so that the Enemy may take an opportunity to stop their March, and to charge them with so much the more advantage in regard that those in the Front and Rear cannot reciprocally come to the Relief of one another. Hence to go off File by File is called *Defiling*.

DEGRADATION, is a Term in Painting, which expresses the lessening and rendring confus'd the Appearance of distant Objects in a Landskip, so as that they shall appear there as they would to an Eye placed at that distance from them.

DEMI-BASTION, is a kind of Fortification that hath only one *Face*, and one *Flank*.

DEMI-GORGE, in Fortification, is half the Gorge or Entrance into the Bastion, not taken directly from Angle to Angle where the Bastion joins to the Cortine, but from the Angle of the Flank to the Certre of the Bastion, or Angle the two Cortins would make were they thus protracted to meet in the Bastion.

DENTILS, in Architecture, is a Member of the *Ionick Cornice*, square and cut out at convenient distances, which gives it the form of a Set of Teeth, and the Original of its Name.

DESSEIN, (in *French*) signifies two things in Painting, either in General the *Design* or Thought that the Painter had about any great Peice; whether there be drawn only the *Out-Lines* (or *Contours*,) or whether he hath proceeded farther to put in the shadows of the Colours, and if there appear much masterly Skill and Judgment in the Thought or Design, we say Design was Great and Noble, *etc.*

It also signifies the Just Measures, the Proportions and External Forms, which those Objects ought to have that are drawn in Imitation of Nature, and in this Sence 'tis one of the Parts of Painting, and may be called a *just Imitation of Nature*.

DIACOUSTICKS, or *Diaphonicks*, is the considerations of the properties of Refracted sound, as it passes though different Mediums.

DIASTYLE, is a sort of Edifice, where the Pillars stand at such a distance one from another, that three Diameters of their thickness are allow'd for *Intercolumniation*.

DIPTERON, in Architecure, is a Name which the Ancients attributed to those Temples which were encompassed with a double Row of Pillars making two Portico's, which they called *Wings*; but we commonly call them *Isles*, from the *French* word *Aisles* of the same signification.

DISTANCE *of the Bastions*, in Fortification, is the side of the *Exterior Polygon*.

DISTEMPER, a Term in Painting for the working up of Colours with something else besides bare Water or Oyl; for if the Colours are prepared with the former of these, that kind of Painting is called *Limning*. And if with Oyl, 'tis called *Painting in Oyl*, and simply *Painting*; but if the colours are mixt with Size, Whites of Eggs, or any such proper Glutinous or Unctuous Substance, and not with Oyl, they can say, 'tis done in *Distemper*; as the admirable *Cartons* are at *Hampton Court*.

DOME, in Architecture, is a round Vaulted or Arched Roof of a Church or any such great Building.

DONJON, in Fortification, is generally taken for a large Tower or Redoubt for a Fortress where the Garrison may retreat in Case of Necessity, and Capitulate with greater Advantage.

DORICK *Order of Architecture*, was invented by the *Dorians* a People of Greece, whence the Name. If its Columns are Simple and Plain without *Pilasters*, *Palladio* saith they ought to be 7 Modules, and one half or Eight Modules high. The Intercolumniations are to be a little less than 3 Diameters of the Column: And this manner of Building is by *Vitruvius* called *Diastyle*.

But if the Columns have *Pilasters*, their height, Reckoning Base and Chapiter, must be 17 Modules and ¼, and by the by, though the Module in all the other Pillars be a Diameter divided into 60 equal Parts, yet in this Order the *Module* is to be accounted but as the Semi-Diameter, and is therefore of 30 such Parts, according to *Palladio*.

This Pillar hath a Proper Base, which is the Reason that mostly in the Ancient Buildings you see them without any Base at all, as in *Marcellus* his Temple at *Rome*, *etc*.

But when the Attick Base is given it, it much Augments its Beauty, and then the height of the Base is to be ½ the Diameter of the Column.

The Chapiters height also ought to be half the Diameter of the Pillar at the Base; and the Architecture the same height. The *Frise* is a *Module* and ½ in height, and the *Cornish* is a *Module* and ⅓. The *Triglyph* is one *Module*, and its *Chapiter* the 6th Part of one, the *Metops* or space between *Triglyph* and *Triglyph* is in length the same with the height of the *Triglyph*.

This Order being designed to represent Solidity to us, ought not to be used but in great and massy Buildings; as the out-side of Churches and Publick Places.

The Entablature here is more massy and tall, than in the other Orders, because the Strength of the Column is so much Greater, it being usually ¼ of the Column. The *Cornish* must be without any Foliage or Trimming, and if you allow *Modillions*, they must be Square and Plain.

The Freeze here hath a Regular Ornament which are the *Triglyphs*, the *Metops* or space between which should be exacty Square.

The Architrave of this Pillar hath also a Peculiar Ornament; which are certain Pendulous Drops under the *Triglyphs*, which yet seem as it were to be fastened to it.

DOUCINE, in Architecture, is an Ornament of the highest Part of the *Cornish*, or a moulding cut in form of a Wave, half Convex, and half Concave.

DRAPERY, signifies in Painting or Sculpture the Cloathing of any Humane Figures, and when the Folds of the Garments hang easie and Natural and yet appear strong, we say the Drapery is very Good.

DRAUGHT *Compasses*, are *Compasses* with several moveable Points to draw fine *Draughts* of *Charts*, *Maps*, *Architecture*, *Fortification*, *Dialling*, etc.

DRIP, in Architecture: See *Larmier*.

DROPS, in Architecture, the same with *Guttæ*.

DURABLE FORTIFICATION, see *Fortification*.

DYE, in Architecture, is the middle of the Pedestal, or that part which lies between their Base and their Cornice; being so made because it is frequently made in form of a *Cube* or *Dye*, and is also called *Dado* by the *Italians*.

DYPTERON, or a *Dypterick* Figure or Order of Pillars in Architecture, is where the Temple Edifice, *etc.* is Invironed round with a Twofold Range of Pillars in the Form of a Double Portico. This *Vitruvius* tells us *ch.* 1. Book 3. was the Figure of the wonderful Temple of *Diana* at *Ephesus*, which was above 200 Years in Building, all *Asia* contributing to its Expence. These Columns were all of Marble, and 70 feet high.

E

ECHINUS, in Architecture, is a Member or Ornament, first placed on the top of *Ionick Capitals*; taking its Name from the roughness of the Carving, resembling the prickly Rind of the Chesnut, and not unlike the Thorny Coat of a Hedge-hog.

This Ornament is made use of by Modern Architects, in Cornices of the *Ionick*, *Corinthian*, and *Composit* Orders, and generally set next to the *Abacus*, being carved with Anchors, Darts, and Ovals or Eggs; whence 'tis called *Ove* by the *French*, and *Ovolo* by the *Italians*; but the *English* Workmen commonly call it the *Quarter-Round*.

EMBRASURE, in Architecture, is the Enlargement made in the Walls, to give more Light and greater Convenience to the Windows or Doors of a Building.

EMBRASURES, in Fortification, are the Holes in a Parapet, through which the Cannons are pointed to Fire into the Moat or Field: They are generally 12 Foot distance one from another; every one of them being from 6 to 7 Foot wide without, and about 3 within; their Height above the Plat-form is 3 Foot on that side toward the Cannon, and a Foot and a half on the other side toward the Field, that so the Muzzle may be sunk on occasion, and the Piece brought to shoot low.

EMPATTEMENT, according to some: is the same with *Talus* in *Fortification*; which see.

ENCEINTE, a *French* term, in Fortification, signifying the whole Enclosure, Circumference, or Compass of a Fortified Place, whether composed of Bastions, or not.

ENFILADE, in Fortification, signifies a situation of Ground which discovers a Post, according to the whole length of a right Line; so that it can be scoured with the Cannon, and rendred almost Defenceless: Wherefore to *Enfile*.

ENFILE, or *Enfilade* the Curtain or Rampart, is sweep the whole length of it with the Cannon.

ENTABLATURE, or *Entablement*, a Term in Architecture, signifying the *Architrave Frise* and *Cornice* of a *Pillar*, being in effect the extremity of the Flooring, which is either supported by Pillars, or by a Wall, if there are not Columns.

ENVELOPE, in Fortification, is a Mount of Earth sometimes raised in the Ditch of a Place, and sometimes beyond it, being either in form of a simple *Parapet*, or of a small *Rampart* bordered with a Parapet. Those *Envelopes* are made when one would only cover the weak Places with single Lines, without any design of Advancing toward the Field, which cannot be done but by Works that require a great deal of breadth; such as *Horn-works*, *Half-Moons*, etc.

These *Envelopes* are sometimes called *Sillons*, *Contregardes*, *Conserves*, *Lunettes*, etc.

EPAULE, in Fortification, is the Shoulder of the Bastion; or the Angle of the *Face* and *Flank*; whence that Angle is often called, the Angle of the *Epaule*.

EPAULMENT, in Fortification, is a *Side-work* made either of Earth thrown up, of Bags of Earth, Gabions, or of Fascines and Earth, of which latter make, the *Epaulments* of the *Places of Arms* for the Cavalry behind the Trenches are.

Sometimes the word EPAULMENT is used for a *Demi-Bastion*; and sometimes it signifies a *Square Orillon*; which is a Mass of Earth almost

Square, faced and lined with a Wall, and design'd to cover the Cannon of a *Cazemate*.

EPISTYLE, in Architecture, is a Mass of Stone, or piece of Timber laid upon the Capital of a Pillar.

The Ancient *Grecians* frequently made use of this Word to signifie that which we call the *Architrave*.

'Tis the first Member of the *Entablature*, and is usually broken into 2 or 3 Divisions, which the Architects call *Fasciæ*, Swathes, Fillets, Bands or Lists.

ESCALADE, or *Scalade*, is a furious Attack upon a Wall, or a Rampart, carried on with Ladders to mount up upon it, without going on in form, breaking Ground, or carrying on of Works to secure the Men.

ESPAULEMENT, the same with *Epaulment*, is a Work in Fortification made on the side of a Bastion, of either Earth thrown up, Gabions, Fascines, *etc.* And those Epaulements which are for the Places of Arms for the Cavalry behind the Trenches, are usually made only of Fascines and Earth. This Word signifies also sometimes a *Demi-Bastion*; and sometimes 'tis used for a Square Orillon, or a Square Mass of Earth, faced or lined with a Wall, designed to cover the Cannon of a *Cazemate*.

ESPAUL, or *Epaule*; (which see); in Fortification, the same with the Shoulder of a Bastion, or the *Angle of the Shoulder*.

ESPLENADE, a Term in Fortification, the same with the Glacis of the Counterscarpe originally; but now 'tis usually taken for the empty Space between the Glacis of a Cittadel, and the first Houses of the Town.

ETCHING, is a way used in making Prints, by drawing with a Needle upon a Copper-plate covered over with a Ground of Wax, *etc.* and well black'd with the Smoak of a Link, that it may take off the Figure of the Drawing or Print, which having its back-side Tinctured with White-Lead will by running over the Stroaken-out Lines with a *Stift*, impress the exact Figure on the Black or Red Ground; which Figure is afterwards with Needles, drawn deeper quite thro' the Ground, and all the *Shadows* and *Hatchings* put in; and then a Wax-Border being made all round the Plate, there is poured on a sufficient Quantity of well-tempered *Aqua fortis*, which insinuating into the Stroaks, and made by the Needles in the *Ground*, eats the Figure of the Print or *Drawing*, into the Copper-Plate. There is no certain Time in which this is done, but usually the *Aqua fortis* will eat deep enough in about half an Hour. *Brown's Ars Pictoria*.

EURITHMY, in Architecture, signifies the exact Proportion between all the Parts of a Building.

EUSTYLE, in Architecture, is a kind of Edifice where the Pillars are placed at a most convenient distance one from another; the *Intercolumniations* being all just two Diameters and a quarter of the Pillar, except those in the middle of the Face before and behind, which are in distance three Diameters.

EYE, in Architecture, is the middle of the *Voluta* of the *Ionick* Chapiter, which is cut in the Form of a little Rose.

F

FACE, in Architecture, is a flat Member which hath a great Breadth, and small Projecture; as in Architraves, *etc.* It is also taken for the Front, or Exterior part of a great Building which immediately presents it self to view.

FACE *of a Bastion, or of the Bulwark,* is the most advanced part of a *Bastion* toward the Field, or the Distance comprehended between the Angle of the Shoulder, and the flanked Angle.

FACE *of a Place*, is the *Curtain* together with the two Flanks raised above it, and the two Faces of the Bastion that look toward one another, and flanking the Angle of the *Tenacle.*

FALSE-BRAYE, in Fortification, is a small Mount of Earth four Fathom wide, erected on the Level round the Foot of the Rampart, on that side towards the Field, and separated by its Parapet from the *Berme*, and the side of the Moat. It is made use of to fire upon the Enemy, when he is already so far advanced, that you cannot force him back from off the Parapet of the Body of the Place. And also to receive the Ruines which the Cannons make in the Body of the Place.

FASCIA, a Term in Architecture, as the *Fascia's* of the Architrave; which are the three Bands of which it is composed.

Vitruvius admits no *Fascia's* in the *Tuscan* or *Dorick* Orders; but is not imitated by any in that particular Circumstance.

FASCINES, or Faggots, in Fortification, are small Branches of Trees, or Bavins bound up in Bundles; which being mixt with Earth, serve to fill up Ditches, to make the Parapets of Trenches, *etc.* Some of them are dipt in Melted Pitch or Tar, and being set on Fire, serve to burn the Enemies Lodgments or other Works.

FASTIGIUM, the same with *Fronton*; which see.

FILET, a little Member in Architecture which appears in the Ornaments and Mouldings, and is otherwise called *Listel.*

FLANK, in Fortification, is that Part of the Bastion which reaches from the Curtain to the Face, and defends the opposite Face, the Flank and the Curtain.

There is also the Oblique, or Second Flank, which is that part of the Curtain, where they can see to Scour the Face of the opposite Bastion; and is the Distance between the Lines *Razant* and *Fitchant*.

The *Low*, *Covered*, or *Retired Flank*, is the Plat-form of the Casemate, which lies hid in the Bastion.

Any Fortification, which hath no Defence right forwards, is on that Account faulty and defective; and to render it compleat, one part must be made *to Flank* another. The Curtain therefore is the strongest part of any fortified Place, because it is *flanked* by the two Flanks at its Ends.

FLANK, in Fortification, is that part of a Bastion, which lies between its Face, and the Curtain; and which serves not only to defend the Curtain, but also the Face of the opposite Bastion.

FLANK *of the Courtine*, or *Second Flank*, is that part of the Courtine between the Flank, and the Point where the *Fichant Line of Defence* terminates.

To FLANK, in general, is to discover and fire upon the Side of any Place; but to

FLANK *a Place*, is to dispose a Bastion, or other like Work, in such a manner that there shall be no part of it but what is defended; so as you may from thence Play upon Front and Rear: For any Fortification that hath no Defence, but just right forwards, is faulty; and to render it compleat, one Part ought to be made to Flank the other: Hence the Curtain is always the strongest part of any Place, because 'tis *flanked* at each end.

Fichant FLANK, is that from whence a Cannon playing fixeth its Bullets directly in the Face of the opposite Bastion.

FLANK *Rasant*, is the Point from whence the Line of Defence begins, from the Conjunction of which with the Curtain, the Shot only raceth the Face of the next Bastion, which happens when the Face cannot be discovered but from the Flank alone.

Retired FLANK, or the *Lower* or *Covert Flank*, is that the exteriour part thereof advanceth to secure the innermost, which advanced part, if it be rounded, is called the *Orillon*; so that this *Flank Retire*, as the *French* call it, is only the Platform of the *Cazemate*, which lies hid in the *Bastion*.

Simple FLANKS, are Lines which go from the *Angle of the Shoulder* to the *Curtain*, and whose principle Function is the Defence of the Moat and Place.

FLANKT *Angle*, is the Angle formed by the two Faces of the Bastion, and so forms the Point of the Bastion.

FLAT *Crown*: See *Corona.*

FOLIAGE, the branched Work in Tapestry or Painting; also a kind of Ornament used in Cornices, Frizes, Chapiters of Pillars, and other Members of Architecture; some of which represent the Leaves of the Herb *Brank-Ursine*, or *Bears-foot*; and others those of divers sorts of Trees, as the *Oak*, *Laurel*, *Olive*, *Vine*, *etc.*

FORT, is a Castle, or a Place of small extent fortified either by Art of Nature.

FORTIFICATION, or *Military Architecture*, is an Art shewing how to Fortifie a Place with *Ramparts*, *Parapets*, *Moats*, and other *Bulwarks*, to the end, that a small Number of Men within, may be able to defend themselves for a considerable time, against the Assaults of a Numerous Army without; so that the Enemy in Attacking them, must of necessity suffer great loss.

Fortification is either *Regular* or *Irregular*, and with respect to Time, may be distinguished into *Durable* and *Temporary.*

Regular Fortification, is that which is built on a *Regular Polygon*, the Sides and Angles whereof are all equal; being commonly about a Musket-shot one from another.

Irregular Fortification, is that where the Sides and Angles are not all uniform, equi-distant, nor equal one to another.

Durable Fortification, is that which is raised to continue a long while.

Temporary Fortification, is that which is erected upon an Emergent occasion, for a little time; such are all sorts of Works cast up for the seizing or maintaining of a Post or Passage, as also *Circumvallations*, *Contravallations*, *Redoubts*, *Trenches*, *Batteries*, *etc.*

***The principal Maxims of* Fortification, *are these*:**

1. Every Place within the *Fortification* ought to be Flank'd; that is, seen sidewise, or defensibe from the other Parts, so that there may be no place in which an Enemy can lodge himself undiscovered by those that are within, and that both from the Front, the Sides, even from behind if possible.

2. The Fortress ought to command all Places round about; and therefore all the Out-works must be lower than the Body of the Place.

3. The Works that are most remote from the Centre of the Place, ought always to be open to those that are more mean.

4. The *Angle Flanquant*, or the Point of the *Bastion*, ought to be at least of 70 Degrees, or as some say, (Mr. *Vauban*) not more than 100, or less than 60.

5. The *Angle of the Courtine*, ought never to be less than 90, or greater than 100 Degrees, because if it be larger, 'tis too much subject to the View of the Enemy.

6. The greater the *Flank* and *Demigorge* is in Proportion to other things, the better; because there is both more Room to retrench in, and also because there may be made retiring Flanks, which add very much to the Strength of a Place.

7. The *Line of Defence* ought never to exceed Point-blank Musket-shot: which is about 120 or 125 fathoms.

8. The *Bastions* that are not too little, nor yet too excessively big, are to be preferred before others; and the Angle of a Bastion should not exceed 100, nor be less than 60 Degrees.

9. The greater the *Angle* that is made by the *outward Polygon* and the Face shall be, the greater is the Defence of the Face.

10. Whatsoever encloses a durable Fortification, must be either Flank, Face or Curtain; and built so well, that the first discharge of the Cannon may not be able to pierce thro' it.

11. 'Tis impossible, to fortifie a Triangle after the common way, because the Angle of the Gorge is always less than 90 Degrees.

12. The Acuter the Angle at the Centre is, the Place is by so much the stronger, because it will have the more Sides.

13. In a *Regular Fortification*, the Face must never be less than half the Curtain; and the Faces of the Bastion ought to be defended by the small shot of the opposite Flanks.

14. *Dry Trenches* are preferable to those filled with Water, especially in great Places, where Sallies, Retreats and Succours are frequently necessary; but in small Fortresses, *Water-Trenches* that cannot be drained, are best, because there is no need of Sallies, Succours, *etc.*

FORTIN, or *Field-Forts*, are Sconces, or little Fortresses, whose *Flank'd Angles* are generally distant one from another 120 Fathom, but their Extent and Figure are different, according to the Situation or Nature of the Ground; some of them having whole Bastions, and others only Demi-Bastions. They are made use of only for a Time, either to defend the Lines of *Circumvallation*, or to guard some Passage or dangerous Post.

FRAISES, in Fortification, are pointed Stakes fixed in Bulwarks made of Earth, on the one side of the Rampart a little below the Parapet. These Stakes being from 7 to 8 Foot long, are driven in almost half way into the Earth, and present their Points somewhat sloaping toward the Field. They serve to prevent Scalades and Desertion.

FREEZE, a Term in Architecture: See *Frize*.

FRESCO, in Architecture, to *Paint in Fresco*, is a sort of Painting which is made upon the Plaistering of an Edifice before it be dry.

FRIZE, or *Freeze*, in Architecture, is that round part of the Entablature which is between the *Architrave* and the *Cornice*. This the *French* call the *Gorgerin* or *Collier*.

FRONT, in Perspective, is the Orthographical Projection of an Object upon a Parallel Plane.

FRONT, in Fortification, is what the *French* call *Tenaille de Place*, and the Face of a Place. It is that which is comprehended between the Points of any two neighbouring Bastions, *viz*. The Curtain, the two Flanks which are raised upon the Curtain, and the two Faces of the Bastions which look towards one another.

FRONTON, is a Part or Member in Architecture, which serves to compose an Ornament raised over Doors, Cross-works, Nitches, *etc.* sometimes making a Triangle, and sometimes part of a Circle. It is also called *Fastigium* by *Vitruvius*, and *Pedisment* by the *French*.

FUSAROLE, is a small round Member in Architecture, cut in form of a Collar, with somewhat long Beads under *Echina's*, or Quarter Round of Pillars of the *Dorick*, *Ionick*, and *Composite Orders*.

G

GABIONS, a Term in Fortification, signifying Baskets made of Osier-Twigs, equally wide at the top and bottom, about four Foot in Diameter, and from five to six high; which being filled with Earth, are sometimes used as Merlons for the Batteries, and sometimes as a Parapet for the Lines of Approach, when it is requisite to carry on the Attacks through a Stony or Rocky Ground, and to advance them with extraordinary Vigour. They serve also to make Lodgments in some Posts, and to secure other Places from the Shot of the Enemies, who nevertheless endeavour to set the Gabions on Fire with pitch'd Faggots to render them useless.

GALLERY, in Fortification, is a covered Walk, the Sides whereof are Musket-proof, consisting of a double row of Planks lined with Plates of Iron; the top being sometimes cover'd with Earth of Turf to hinder the Effect of the Artificial Fire of the Besieged. These Galleries are frequently made use of in the Moat already filled with Faggots and Bavins, to the end that the Miner may approach safe to the Face of the Bastion, when the Artillery of the opposite Flank is dismounted.

GLACIS, a sloaping Bank in Fortification: It signifies a very gentle steepness, but is more especially taken for that which rangeth from the Parapet of the *Cover'd-way*, to the Level on the side of the Field.

GORGE, *Gulla*, or *Neck*, in Architecture, is the narrowest part of the *Tuscan* and *Dorick* Capitals, lying between the *Astragal*, above the Shaft of the Pillar, and the Annulets. It is also, a kind of Concave-Moulding, larger, but not so deep as a Scotia, which serves for Compartments, Chambrances, *etc*.

GORGE, in Fortification, is the Entrance of the Platform of any Work.

GORGE, in all other Out-Works, is the Interval betwixt their Wings on the side of the great Ditch: But it ought to be observed, that all the *Gorges* are destitute of *Parapets*, because if there were any, the Besiegers having taken Possession of a Work, might make use thereof, to defend themselves from the Shot of the Place; so that they are only fortified with *Pallisadoes* to prevent a Surprise.

GORGE *of a Bastion*, is nothing else but the prolonging of the *Curtains* from their Angle with *Flank*, to the Centre of the *Bastion* where they meet: But when the Bastion is flat, its *Gorge* is a right Line which terminates the Distance comprehended between two Flanks.

GORGE *of the Ravelin*, or of a *Half-Moon*, is the Space contained between the Extremities of the two Faces on the side of the Place.

GOTHICK, in Architecture, is an Order so far different from the Ancient Proportions and Ornaments, that its Columns are either too massy in a form of vast Pillars, or as slender as Poles, having Capitals without any certain Dimensions, carved with the Leaves of *Brank-Usine*, *Thistles*, *Coleworts*, *etc*.

GULBE, in Architecture, the same with *Gorge*.

GUTTÆ, or *Drops*, in Architecture, are certain Parts in form of little Bells, which to the Number of Six, are put below every *Triglyph*, in the *Architrave* of the *Dorick* Order. They are so called by Architects from their Shape, representing the *Drops* of Water, which having run along the *Triglyphs*, still hang under the closure between the Pillars.

GUVE *de Ronde*, a Term in Fortification, signifying the same as a simple or a *Single Tenaille*. See *Tenaille*.

H

HALF-MOON, in Fortification, is an Out-work, that hath only two Faces forming together a Saliant-Angle, which is flanked by some part of the Place, and of the other Bastions.

These *Half-Moons* are sometimes raised before the Curtain when the Ditch is a little wider than it ought to be; in which sense, 'tis much the same with a Ravelin; only the Gorge of an Half-Moon is made bending in, like a Bow or Crescent, and is most times used to cover the Point of a Bastion, whereas Ravelins are placed before the Curtain; but they are defective, as being ill flanked.

HARMONY, is an agreeable or pleasing Union between two or more Sounds, continuing together at the same time.

Harmony is naturaly produced by *Consonances*, but yet Art has discover'd the way to make it yet more agreeable by the mixture of *Dissonances*.

HEPATAGON, in Fortification, is taken for a Place that hath seven *Bastions* for its Defence.

HERSE, in Fortification, is a *Lattice*, or *Portcullice*, made in the form of a *Harrow*, and beset with many Iron Spikes. It is usually hung by a Cord fastned to a Moulinet, which is cut in case of Surprise, or when the first Gate is broken with a *Petard*, to the end that the *Herse* may fall, and stop up the Passage of the Gate, or other Entrance of a Fortress. These *Herses* are also often laid in the Roads to incommode the March as well of the Horse, as of the Infantry.

HERSILLON, in Fortification, is a Plank stuck with Iron Spikes, for the same use as the *Herse*.

HEXASTYLE, an Ancient Building which had six columns in the Face before, and six also behind, and is the same with the *Pseudodipteron*.

HOLLOW-TOWER, in Fortification, is a Rounding made of the Remainder of two Brisures, to joyn the Curtain to the *Orillon*, where the Small-shot are plaid, that they may not be so much exposed to the View of the Enemy.

HORN-WORK, in Fortification, is an Out-work which advanceth toward the Field, carrying in the fore-part, or its Head, two Demi-Bastions, in form of *Horns*. These Horns, Epaulments or Shouldrings, being joyn'd by a Curtain, shut up on the side by two Wings parallel one to another, are terminated at the Gorge of the Work, and so present themselves to the Enemy.

HURDLES, or *Clays*, in Fortification, are made of thick and small Twigs of Willows, or Osiers, being 5 or 6 Foot high, and from 3 to 4 Foot broad. They are interwoven very close together, and usually laden with Earth, that they may serve to render Batteries firm, or to consolidate the Passage over Muddy Ditches; or to cover Traverses and Lodgments for the Defence of the Workmen, against the Artificial Fires or Stones that may be cast upon them.

HYPERTHYRON, in Architecture, is a large Table usually placed over Gates or Doors of the Dorick Order, above the Chambranle, in form of a *Frize*.

HYPOTRACHELION, in Architecture, is the Top or Neck of a Pillar, or the most slender part of it which toucheth the Capital.

It is taken by some, for that part of the *Tuscan* and *Dorick* Capitals, which lies between the *Echinus* and the *Astragal*, and is otherwise called, the *Collar*, *Gorge*, or *Frize of the Chapiter*.

I

ICHNOGRAPHY, in Fortification, is in like manner the Plane or Representation of the length and breadth of a Fortress, the distinct parts of which are mark'd out either upon the Ground it self, or upon Paper.

ICHNOGRAPHY, in Perspective, is the View of any thing cut off by a Plane Parallel to the Horizon, just at the Base or Bottom of it. And in Architecture is taken for the Geometrical Plan, or Plat-form of an Edifice, or the Ground-plot of a House or Building delineated upon Paper, describing the Form of the several Apartments, Rooms, Windows, Chimneys, *etc*. And this is properly the Work of the Master Architect or Surveyor, being indeed the most abstruse and difficult of any.

INTERCOLUMNIATION, is the Space or Distance between the Pillars of any Building.

INTERCOLUMNS, or *Intercolumniations* in Architecture, are the Spaces between Column and Column in any Portico or great Piazza, *etc*.

INTERIOR TALUS. See *Talus*.

INWARD FLANKING-ANGLE, in Fortification, is made by the *Courtine* and the *Razant Flanking-Line* of *Defence*.

IONICK *Order of Architecture*, is the Form of a Column or Pillar invented by the *Ionians* in Ancient *Greece*, by way of Improvement of the Beauty of the *Dorick Order*: Which as it was taken from the Figure of a Robust Man's Body, and Designed to represent Solidity and Strength: So the *Ionians* having more regard to Beauty, chose an Order of a more Elegant Proportion, which occasioned this Order to be called, *The Feminine Order*, and soon after it, sprung up that of the *Caryatides*.

The Proportions of this Pillar, as they are taken from the Famous one in the Temple of *Fortuna Virilis* at *Rome*, now the Chuch of St. *Mary* the *Egyptian*; are these.

The entire Order from the Superficies of the *Area* to the *Cornice*, are 22 Modules or 11 Diameters.

The Column with its Base and Capital, contains 18 Modules.

The *Entablature* (*i.e.* the *Architrave-Freeze* and *Cornice*) contain four Modules.

The *Voluta* of the Capital, is of an Oval Form.

The Columns in this Order are often hollow'd and furrow'd with 24 Gutters, and sometimes 'tis done only to the third part of the Column, reckoning from the Bottom; and then that third part hath its Gutters filled with little Rods or Battoons, all the part of the Hollow above being left empty.

ISLES, in Architecture, are sides or Wings of a Building.

J

JET D'EAU, is the *French* Word for a Pipe of a Fountain which casts up the Water any considerable height into the Air.

Mr. *Mariotte* in his Treatise *du Mouvement des Eaux*, etc. saith, That a *Jet d'Eau* will never rise so high as its Reservatory, but always falls short of it by a Space which is in a subduplicate Ratio of that height; and this he proves by several Experiments.

He saith also, That if a greater branches out in many smaller ones, distributed to different Jets, the Square of the Diameter of the Main Pipe must be proportioned to the Sum of all the Expences of its Branches; and particularly saith, That if the Reservatory be 52 Foot high, and the *Ajutage* half an Inch in Diameter, the Pipe ought to be 3 Inches in Diameter.

L

LACUNAR, in Architecture, is an Arched Roof or Cieling, more especially the Planking or Flooring above the *Portico's*.

LARMIER, a flat square Member in *Architecture*, which is placed on the Cornice below the *Cymatium*, and jets out farthest; being so called from its Use, which is to disperse the Water, and to cause it fall at a Distance from the Wall, Drop by Drop, or as it were by Tears: For *Larme* in *French* signifies a Tear. See *Corona*.

LINE, in Fortification, is that which is drawn from one Point to another, in delineating a Plane upon Paper. But in the Field it is sometimes taken for a Ditch bordered with its Parapet, and sometimes for a Row of *Gabions*, or Sacks or Earth, extended in length on the Ground, to serve as a Shelter against the Enemies Fire. Thus they say, when the Trenches were carried on within 30 Paces of the *Glacis*, we drew two Lines, one on the Right Hand, the other on the Left, for a Place of Arms.

LINE *Capital*, is that which is drawn from the *Angle of the Gorge* to the *Angle of the Bastion*.

LINE *Cogritel*, is that which is drawn from the *Angle of the Centre* to that of the *Bastion*.

LINE *of Defence*, is that which represents the Course of the Bullet of any sort of Fire-arms, more especially of a Musquet-Ball, according to the Situation which it ought to have to defend the Face of the Bastion.

LINE *of Defence Fix'd or Fichant*, is that which is drawn from the *Angle of the Curtain* to the *flank'd Angle* of the opposite *Bastions*; nevertheless without touching the Face of the Bastion. This must never exceed 800 Feet, which they reckon the Distance a Musquet-Ball will do execution.

LINE of *Defence Razant*, is that which being drawn from a certain Point of its Curtain, raseth the Face of the opposite Bastion. This is called also the *Line of Defence Stringent* or *Flanking*.

LINE of *Approach* or of *Attack*, signifies the Work which the Besiegers carry on under *Covert*, to gain the Moat, and the Body of the Place.

LINE *of Circumvallation*, is a *Line* or *Trench* cut by the Besiegers within Cannon-shot of the Place, which rangeth round their Camp, and secures its Quarters against the Relief of the Besieged.

LINE *of Contravallation*, is a Ditch border'd with a Parapet, which serves to cover the Besiegers on the side of the Place, and to stop the Sallies of the Garrison.

LINES *within side*, are the Moats toward the Place, to prevent the like Sallies.

LINES *without side*, are the Moats towards the Field, to hinder Relief.

LINES *of Communication*, are those that run from one Work to another. But the *Line of Communication*, more especially so called, is a continued Trench with which a *Circumvallation* or *Contravallation* is surrounded, and which maintains a *Communication* with all its *Forts*, *Redoubts*, and *Tenables*.

LINE *of the Base*, is a Right Line which joins the Points of the two nearest *Bastions*.

To LINE *a Work*, is to strengthen a Rampart with a firm Wall, or to encompass a Parapet or Moat with good Turf, *etc*.

LIST, in Architecture, is a strait upright Ring which runs round the lower part of any of the Columns just above the *Tore*, and next to the Shaft or Body of the Pillar.

LISTEL, a small Band, or a kind of Rule in the Mouldings of Architecture. Also the Space between the Channelings of Pillars.

LIZIERE, a Term in Fortification, the same with *Berme*; which see.

LODGMENT, in Military Affairs, is sometimes an Encampment made by an Army; or oftener, it is a Retrenchment dug for a Covert or Shelter, when the Counterscarp or some other Post is gain'd: It is also taken for the Place where the Soldiers quarter among the Burghers, either in Huts, Baraques, or Tents.

LODGMENT *of an Attack*, is a Work cast up by the Besiegers, during their Approaches in a dangerous Post, where it's absolutely necessary to secure themselves against the Enemies Fire; as in a *Covert-way,* in a *Breach*, in the bottom of the Moat, or elsewhere. This Lodgment consists of all the Materials that are capable to make Resistance, *viz. Barrels* and *Gabions* of Earth, *Pallisadoes*, *Woolpacks*, *Mantelets*, *Faggots*, etc.

LUNETTES, in Fortification, are *Envelopes*, *Countergardes*, or Mounts of Earth cast up before the Curtain, about five Fathom in breadth, whereof the Parapet takes up three. They are usually made in Ditches full of Water, and serve to the same purpose as *False Braies*. These *Lunettes* are compos'd of two Faces, which form a Re-entring Angle; and their Platform being only Twelve Foot wide, is a little raised above the Level of the Water, and hath a Parapet three Fathom thick.

M

MADRIER, in Fortification, is a thick Plank arm'd with Plates of Iron, and having a Concavity sufficient to receive the Mouth of the Petard when charged, with which it is applied against a Gate, or any thing else that you design to break down. This Term is also appropriated to certain flat Beams, which are fix'd at the bottom of a Moat, to support a Wall. There are also *Madriers* lined with Tin, which are cover'd with Earth, to serve as a Defence against Artificial Fires.

MANNER. A Word now much in use, which we have borrowed from the French *Maniere*. In Painting it signifies the Usage, Way, *Mode*, or *Manner* any Painter hath acquired, not only in the Management of his Hand or Pencil, but also as to his Observance of the Three Principal Parts of Painting, *Invention*, *Design*, and *Colour*: And according as any one hath gotten a Habit or peculiar Way of Painting, we say he hath such a *Manner*. If it be agreeable to the Rules of this Art, Natural, Strong, Easie, and Duly Proportioned, we say 'tis a *Good manner*; and if the Contrary, a *Bad manner*. If he imitate any Famous Ancient Painter, as *Michael Angelo*, *Raphael*, etc. we say *He paints after their manner*. So 'tis also in Sculpture; and now adays in *Singing*, or *Playing* on any Instrument: When we would express our Approbation of any one's Way of *Singing* or *Playing*, we say *He hath a very Good manner*.

MANTELETS, in Fortification, are a kind of movable Penthouses, and are made of pieces of Timber saw'd into Planks; which being about three Inches thick, are nail'd one over another to the Height of almost six Foot. They are generally cased with Tin, and set upon little Wheels, so that in a Siege they may be driven before the Pioneers, and serve as Blinds, to shelter them from the Enemies Small-shot. There are also other sorts of *Mantelets*, cover'd on the top, whereof the Miners make use to approach the Walls of a Town or Castle.

MERLON, in Fortification, is that Part of the Parapet which lies betwixt two Embrasures, being from 8 to 9 Foot long on the side of the Cannon, and 6 on the side of the Field; as also 6 Foot high, and 18 thick.

METOPA, in Architecture, is the Interval or Space between every Triglyph in the Frize of the *Dorick* Order. The Ancients used to adorn these Parts with Carved Works or Paintings, representing the Heads of Oxen, Vessels, Basons, and divers other Instruments that were used in their Sacrifices.

MILITARY *Architecture*, the same with *Fortification.*

MINE, in Fortification, is a Hole dug or made by a Pioneer under the Rampart, or under the Face of the Bastion, whereto there are several oblique and winding Passages: When it is finished, divers Barrels of Powder are placed therein, together with a Train or *Saucidge*; and the Quantity of Powder is proportion'd to the Height and Weight of the Body which is to be blown up. There are also *Mines* sprung in the Field, which are called *Fougades.*

The Ally or Passage of a *Mine* is usually about Four Foot square, at the end of which is the *Chamber* of the *Mine*, as they call it. The farther it is carried on, the more it is subject to be discovered by the Enemy; therefore 'tis best not to aim at Mining too far, and to make a new one where the former takes no effect.

MOAT, in Fortification, is a hollow Space or ditch dug round a Town or Fortress which is to be defended; whereof the Length and Breadth often depends upon the Nature of the Soil, according as it is Marshy or Rocky. But *Moats* in general may be from 16 to 22 Fathom broad, and from 15 to 25 Foot deep.

Dry Moat, is that which is destitute of Water, and ought to be deeper than one that is full of Water.

Lined Moat, is that whose Scarp and Counterscarp are cas'd with a Wall of Masons Work lying in *Talus* or a-sloap.

Flat-bottom'd Moat, is that which hath no sloaping, its Corners being somewhat rounded. All Moats must be well flanked, and in general so wide, as that no Ladder, Tree, *etc*. can reach across it. If the Ditch be dry, or has but little Water, there is usually another small Trench cut quite along the middle of it.

MODEL, *in Architecture*: See *Module.*

MODILLONS, or *Modillions*, in Architecture, are little Brackets which are often set under the Cornices, more especially in the *Corinthian* and *Composit* Order, and serve to support the Projecture of the *Larmier* or *Drip*. The Word comes from the Italian *Modiglion*, signifying a little Model or Measure; but this Part must be distinguish'd from the great Model, which is the Diameter of the Pillar: For as the Proportion of an Edifice in general depends on the Diameter of the Pillar, so the Size and

Number of the *Modillons*, as also the Interval between them ought to have due Relation to the whole Fabrick.

MODULE, or *Model*, in Architecture, is a certain Measure invented by *Vignola*, and made use of to regulate the Proportion of the whole Building. It is generally half the Diameter of a Pillar at the lower end, in the *Tuscan* and in the *Dorick* Order; but in others, the whole Diameter. This Diameter is divided into 12 equal Parts; and into 18 for the *Ionick*, *Corinthian*, and *Composite* Orders. And this *Module* or *Model* is a kind of Universal Measure, which helps us to get rid of the great Uncertainties there are in the Feet and Inches of divers Nations, and at diverse Times.

MOINEAU, is a Name the *French*, and some Modern Writers of Fortification, give to a little *Plat-Bastion*, which is raised before a Curtain that is too long, and which hath two other Bastions at the ends of it; for they being out of Musket-shot one of the other, must be defended by some such thing as this *Moineau* or *Plat-Bastion*. Sometimes the *Moineau* joins to the Curtain, and sometimes is disjoined from it by a Moat.

MONOTRIGLYPH, a Term in Architecture, signifying the Space of one *Triglyph* between two *Pillasters*, or two *Columns*.

MOSAICK *Work*, was anciently used only in Pavements, and began in *Rome* about *Sylla's* time; who made a Pavement of *Mosaick* Work at *Preneste* in the Temple of *Fortune*, about 170 Years before Christ. It was called *Lithestroton*, in *Greek* only a Stone Pavement; but it is understood of Figures made or represented by Inlaying of small Stones, or rather pieces of Stone of different Colours; by the variety of which, many Curious Figures may be wrought. Afterward it came to be in Fashion for the Insides of the Walls of Rooms; and now adays they work it with Shells, or small Pieces of Glass variously Coloured and Figured. 'Tis a very Pleasant and a lasting Ornament, when 'tis curiously and excellently done.

MOYENAU (a *French* Term) in Fortification, is a small flat Bastion, commonly placed in the middle of an over-long Curtain, by which the Bastions at the Extremities are not well defended from the Small-shot, by reason of their Distance; so that this Work is proper for placing in it a Body of Musqueteers to fire upon the Enemy from all sides.

MUSKET-*Baskets*, in Fortification, are Baskets of about a Foot and half high, and 8 or 10 Inches Diameter at the Bottom, and a full Foot at the top: They are filled with Earth, and are set on low Parapets or Breast-works, or on such as are beaten down, that the Musketers may fire between them at the Enemy, and yet be tolerably well secured against their Fire.

MUTULE, in Architecture, is a kind of Square *Modillon*, set under the Cornice of the *Dorick* Order, and so called from the Word *Mutilus*, Maim'd, or Imperfect, because they represent the ends of the Rafters which are crooked or bent; in like manner, as the Beams, or Joints are represented by the *Triglyphs* in the *Frize* of the same *Order*.

N

NICHE, in Architecture, is a Cavity left designedly in the Wall of a Building, to place a Statue in.

NORMAL, the same with Perpendicular, or at Right Angles, and 'tis usually spoken of a Line or a Plane that Intersects another Perpendicularly.

NUCLEUS, in Architecture, is the middle part of the Flooring of the Ancients, consisting of Cement, which they put betwixt a Lay or Bed of Pebbles cemented with Mortar made of Lime and Sand.

O

OBELISKE, is a magnificent high Piece of Solid Marble, or other fine Stone; having usually Four Faces, and lessening upwards by Degrees, till it ends in a Point like a Pyramid.

OCTOSTYLE, in *Architecture,* is the Face of an Edifice adorned with Eight Columns.

OPEN *Flank*, in Fortification, is that part of the *Flank* which is covered by the *Shoulder* or *Orillion*.

ORDER in *Architecture*, is a Rule for the Proportion to be observed in the erecting of Pillars, and for the form of certain parts appertaining to be them.

So Buildings are said to be of several *Orders*, when the Proportion between the thickness of the Columns, and their height, together with all things requisite thereto are different.

There are Five Principal Orders *of* Architecture, *viz. The* Tuscan, Dorick, Ionick, Corinthian, *and the* Composit.

The *Tuscan* is the most Simple, and the most destitute of Ornaments, so that it is seldom used except in Vaults, in some Rustick Edifices, vast Piles of Buildings, as *Amphitheatres, etc.*

The *Tuscan* Pillar with its Base and Chapiter generally hath for its height seven Diameters of its thickness taken below, and the top ought to be diminished one quarter of its Diameter: The Pedestal is very plain, and only one Model high.

The *Dorick Order*, hath its Column eight Diameters high, and ought to have no Ornament neither in its Chapiter nor Base. The *Astragal* and *Listel* below the Capital, which is half a Diameter high, constituting part of the Shank or Body of this Pillar.

The *Ionick Order*, when first invented had its Column only eight Models high; but the Ancients designing to make it more beautiful, augmented the height of the Pillars, and added a Base not used before; so that with its Chapiter and Base it contains nine Diameters of its thickness taken below: the Pedestal is two Diameters, and about two thirds high; and the Capital is chiefly composed of *Voluta's* or Scrolls, which render it different from the other *Orders*: The *Ionick* Pillars are also usually channelled with 24 Flutes.

The *Corinthian Order*, is the finest and richest of all; the length of its Columns with their Bases and Chapiters, is commonly about nine and a half or ten Diameters; and the Capitals are adorn'd with two Rows of Leaves, and eight *Voluta's* that support the *Abacus*.

The *Composit* or *Roman Order*, hath the Capital of its Pillar composed of two rows of Leaves like the *Corinthian*, and of the *Voluta's* or Scrolls of the *Ionick*. These Columns are usually ten Diameters high, altogether like the *Corinthian* in all its Dimensions and Numbers, except the Capital, which hath only four *Voluta's* taking up the whole Space, which is filled both by the *Voluta's* and Stems or Stalks in the *Corinthian*.

To these Five Orders, *may be added also.*

The *Attick*, which is a small *Order* of *Pillasters* of the shortest Proportion, having a Cornice raised after the manner of an Architrave, for its Entablature. As also

The *Gothick Order*, which is so far from the Ancient Proportions and Ornaments, that its Columns are either too Massie in Form of vast Pillars, or as slender as Poles, having Capitals without any certain Dimensions, carved with the Thorny Leaves of *Thistles*, *Coleworts*, *Bearsfoot.*

ORDONANCE, in Architecture, is the giving to all the Parts of an Edifice that just Quantity and Dimensions which they ought to have, according to the Model.

ORILLON, in Fortification, is a small Rounding of Earth lined with a Wall, which is raised on the Shoulder of those Bastions that have Casemates, to cover the Cannon in the Retired Flank, and prevent their being dismounted by the Enemy.

There are also other sorts of *Orillons*, properly called *Shoulderings*, which are almost of a square Figure, and are called *Epaulements.*

ORLE, a Term in Architecture; the same with *Plinth*, which see.

ORNAMENTS, in Architecture, are the *Architraves*, *Frizes*, and *Cornices* of the several *Orders*. But there are also many Ornaments frequently carved in the Mouldings, and on all other Members; as divers sorts of *Leaves*, *Channelings*, *Wreaths*, *Ovals*, *Chapelets*, *Tresecs*, *Festoohs*, *Flowers*, *Roses*, *etc.* The *Ornaments* of the Columns the *French* call *Oeufs*.

ORTEIL, a Term in Fortification; the same with *Berme*, which see.

ORTHOGONIAL, the same with *Right-angled*; and when referred to a Plain Figure, supposes one Leg or Side to stand perpendicular to the other; And when it is spoken of Solids, it supposes their Axis to be perpendicular to the Plane of the Horizon.

ORTHOGRAPHY, in *Mathematicks*, is the true Delineation of the fore-right Plain of any Object.

In *Architecture*, 'tis taken for the Model, Platform, and Delineation of the Front of a House that is to be built and contrived according to the Rules of Geometry; according to which Pattern, the whole Fabrick is erected and finished.

In *Perspective*, the *Orthography* of any Body, or Building, is the foreright side of any Plane; that is, the Side or Plane that lies parallel to a straight Line, that may be imagined to pass thro' the outward Convex Points of the Eyes, continued to a convenient Length. The Word Scenography is used by *Lamy* and Others in the same Sense.

In *Fortification*, it is the *Profil* or Representation of a Fortress, made after such a manner, that the Length, Breadth and Height of its several Parts may be discovered.

OUT-*Works*, in Fortification, are all sorts of Works which are rais'd without the Inclosure of a Place, and serve for its better Defence, and to cover it from the Enemy in the Plain without; as, *Ravelins*, *Half-moons*, *Horn-works*, *Crown-works*, *Counterguards*, *Tenailles*, etc.

It is a General Rule in all Out-works, that if there be several of them one before another to cover one and the self-same *Tenaille* of a Place, the nearer ones must gradually, and one after another, command those which are farthest advanced out into the Campaign; that is, must have higher Ramparts, that so they may over-look and fire upon the Besiegers, when they are Masters of the more outward Works.

The *Gorges* also of all Out-works must always be plain, and without Parapets; lest, when taken, they should serve to secure the Besiegers against the Fire of the Retiring Besieged; wherefore the *Gorges* of Out-works are only Palisado'd, to prevent a Surprize.

OVAL, in Architecture, the same with *Echinus*. Some write it *Ovo*, because of its Figure, being like an Egg. It is placed in the Mouldings of the Cornices, for Ornament; and in a Pillar it is placed next the *Abacus*.

P

PALLIER, a Landing-place in a Stair-Case, or a Step, which being much broader than the other, serves to rest upon.

PALLISADOES, or *Pallisades* in Fortification, are strong Wooden sharp-pointed Stakes, six or seven Inches square, eight Foot long, of which three Foot is in the Ground, set up half a Foot sometimes one above another, with a cross piece of Timber that binds them together. Some of these are also sometimes arm'd with two or three Iron Spikes.

These *Pallisadoes* are usually fixed in the void Spaces without the *Glacis*, near the Bastions and Curtains; and in the Avenues of all such Posts as are liable to be Surpriz'd by the Enemy, or carried by Assault. Sometimes they are driven downright in the Ground, and sometimes stand at an Acute Angle towards the Enemy, that if they should thro' Cords about them to pull them up, they may slip off again.

Pallisadoes are always planted on the *Berme* of Bastions, and at the *Gorges* of Half-Moons and other Out-works: They also *Pallisade* usually the Bottom of the Ditch; and to be sure, the *Parapet* of the *Covert way*: and tho' sometimes they have placed these Pallisades three Foot from the said Parapet outwards towards the Campaign, yet of late they have been planted in the very middle of the Covert way: All Pallisades should stand so close, as to admit between them only the Muzzle of a Musket, or a Pike.

PARAPET, in Fortification, is an Elevation of Earth and Stone upon the Rampart, behind which the Soldiers stand secure from the Enemies Great and Small Shot, and where the Cannon is planted for the Defence of the Town or Fortress. Every *Parapet* having its *Embrasures* and *Merlons*, is about Six Foot high on the side of the Place; and from Four to Five on that toward the Country. So that this Difference of Heights forms a kind of *Glacis* above, from whence the Musketeers mounting the *Banquet* of the *Parapet*, may easily fire into the Moats, or at least upon the Counterscarp. It ought also to be from 18 to 20 Foot thick, if made of Earth; and from 6 to 8 if of Stone. The Earth is much better than Stone, because Stone will fly to pieces when battered, and do mischief.

This Word *Parapet* is also given to any Line that covers men from the Enemies Fire: So there are *Parapets* of Barrels, of Gabions, of Bags filled with Earth, *etc.*

PEDESTAL, in Architecture, is that part which supports a Pillar or Column, having its proper Base and Cornice different, according to the several Orders of Architecture.

The *Tuscan Pedestal*, being the most simple of all, hath only a *Plinth* for its Base, and an *Astragal* crown'd for its *Cornice.*

The *Dorick Pedestal*, (according to *Palladio*) borrowing the *Attick* Base, ought to have for its Height 2⅓ Diameters of the Column taken below. But no *Pedestals* to this *Order* are seen among the Ancient Buildings.

The *Ionick Pedestal* is Two Diameters and about Two thirds high.

The *Corinthian Pedestal* hath the Fourth part of the Column for its Height, being divided into Eight parts; whereof one must be allowed for the *Cymatium*, two others for the Base, and the rest for the Dye or Square.

The *Composit Pedestal* ought to have the third part of the Pillar for its Height.

PEDIEMENT, a Term in Architecture; the same with *Fronton*, which see.

PERAMBULATOR, or *Walking-wheel*, is an Engine made of Wood or Iron, commonly half a Pole in Circumference, with a Movement and a Face divided, like a Clock with a long Rod of Iron, or Steel that goes from the Centre of the Wheel to the Work: There are also two Hands, which (as you drive the Wheel before you) count the Revolution, and from the Composition of the Movement, and by the Division on the Face, shew how many Yards, Poles, Furlongs and Miles you go. The use of this Instrument is to measure Roads, Rivers, and all Level Lands, with great expedition.

PERIPTERE, in Architecture, is a Place encompassed round with Columns, and with a kind of Wings about it; here the Pillars stand without, whereas in the *Peristyle* they stand within.

PERISTYLE, in Architecture, is a Place or Building encompass'd with Pillars, standing round about within the Court: But this Word *Peristyle* is sometimes taken for a row or rank of Columns, as well without as within any Edifce; as in Cloysters and Galleries: Sometimes this was call'd *Antiprostyle*.

PERSICK *Order of Architecture*, is where the Bodies of Men serve instead of Columns to support the *Entablature*, or rather the Columns are in that form.

The Rise of it was this; *Pausanias* having defeated the *Persians*, the *Lacedemonians*, as a Mark of their Victory, erected Trophies of the Arms of their Enemies, and then represented the *Persians* under the Figures of Slaves supporting their *Porches*, *Arches*, or *Houses*.

PERSPECTIVE, is that part of the Mathematicks which gives Rules for the representing of Objects on a plain Superficies, after the same manner as they would appear to our sight, if seen thro' that Plain, it being supposed as transparent as Glass.

***Perspective* is either *Speculative* or *Practical*.**

Speculative Perspective, is the Knowledge of the Reasons of the different Appearance of certain Objects, according to the several Positions of the Eye that beholds them.

Practical Perspective, is the Method of Delineating that which is apparent to our Eyes, or that which our Understanding conceives in the forms that we see Objects.

PERSPECTIVE *Aerial*, is a Proportionable Diminution of the Teints and Colours of a Picture, when the Objects are supposed to be very remote.

PERSPECTIVE *Lineal*, is the Diminution of those Lines in the Plan of a Picture, which are the Representations of other Lines very remote.

PERSPECTIVE *Military*, is when the Eye is supposed to be infinitely remote from the Table or Plain.

PICKET, in Fortification, is sometimes used for a Stake, sharp at one end, to mark out the Ground and Angles of a Fortification, when the Engineer is laying down the *Plan* of it; these are usually pointed with Iron. There are also larger *Pickets*, which are drove into the Earth, to hold together *Fascines* or Faggots, in any Work cast up in haste.

PIED-DROIT, in Architecture, is a Square Pillar, differing from a *Pillaster* in this respect, that it hath no Base nor Capital: It is taken also for part of the Jaumbs of a Door or Window.

PIEDOUCHE, in Architecture, is a little Square Base smoothed, and wrought with Mouldings, which serves to support a *Bust* or Statue drawn half way, or any small Figure in Relief.

PIGMENTS, are such prepared Materials as Painters, Dyers, *etc.* make use of, to impart to Bodies, or to imitate particular Colours. When Glass is stain'd or Coloured, as in painting on Glass, or for the Counterfeiting of Gems or Precious Stones, the *Pigment* is usually of a Metalline or a Mineral Nature.

PILLAR or *Column*, in Architecture, is one of the principal Things in which the Beauty and Proportion of a Fabrick doth consist. A *Pillar* hath Three Parts, the *Pedestal*, the *Shaft* or *Pillar* itself, and the Ornaments: And each of these is again subdivided into Three other Parts; the *Pedestal* hath its *Base*, *Dye*, and *Cornich*; the *Pillar* its *Base*, *Shaft*, and *Capital*; and the *Ornaments* have their *Architrave*, *Frieze*, and *Cornich.*

PILLASTERS, in Architecture, signifie Square Pillars, that usually stand behind Columns to bear up Arches: They have the very same Dimension, Chapiter, and Base, with the Columns, according to their several *Order*. When these *Pillasters* do not stand alone by themselves, they are usually made to jet out of the Wall, a Third part or a Quarter of their Breadth, with respect to the Variety of different Works: So that some of them project out only a Sixth or an Eighth part, and they are generally as broad at top as at bottom.

PLACE *of Arms*, when taken in the General, is a strong City which is pitch'd upon for the Magazine of an Army. But a

PLACE in Fortification usually signifies the Body of a Fortress. And a

PLACE *of Arms in a Garison*, is a large open Spot of Ground in the middle of the City, where the great Streets meet, else between the Ramparts and the Houses, for the Garrison to Rendezvous in, upon any sudden Alarm, or other Occasion. And the

PLACE *of Arms of a Trench*, *or of an Attack*, is a Post near it, shelter'd by a *Parapet* or *Epaulement*, for Horse and Foot to be at their

Arms, to make good the Trenches against the Sallies of the Enemy. These Places of Arms are sometimes covered by a *Rideau* or Rising Ground, or else by a *Cavin* or Deep Valley, which saves the trouble of Fortifying them by means of Parapets, Fascines, Gabions, *etc.* They are always open in the Rear, for their better Communication with the Camp. When the Trench is carried on as far as to the *Glacis*, they make it very wide, that *it* may serve for a Place of Arms. Also

PLACE *of Arms of a Camp*, is a spacious piece of Ground at the Head of the Camp, to draw out the Army in Order of Battle. But the

PLACE *of Arms* of a Troop of Horse, or of a Company of Foot in the Camp, is that Spot of Ground on which the Troops or Company draws out.

PLANCERE, in Architecture, is the Under part of the Roof of the *Corona*; which is the Superior part of the Cornice, between two *Cymatiums*. See those Words. (For Cornice see *Corniche.*)

PLANE, in Fortification, is the Representation of Work in its Height and Breadth.

PLANE *Geometrical*, in Perspective, is a Plain Surface, parallel to the Horizon, placed lower than the Eye; wherein the visible Objects are imagined without any Alteration, except that they are sometimes reduced from a greater to a lesser Size.

PLANE *Horizontal*, in Perspective, is a Plane which is parallel to the Horizon, and which passes thro' the Eye; or hath the Eye supposed to be placed in it.

PLASM, the same with a Mould in which any Metal, or such like running Matter, which will afterwards harden, is cast.

PLAT-*Band*, in Architecture, is a Square Mould which terminates the Architrave of the *Dorick Order*, or the *Fascia* which passeth immediately under the *Triglyphs*, and serves for the same Use in this Order, as the *Cymatium* in the others. It is also the Fascia of the *Chambranles*: And the same Name is also attributed to divers other Members of Architecture, which are desitiute of Ornaments, having only a certain breadth without much Projecture.

PLATFORM, in Architecture, is a row of Beams that support the Timber-work of a Roof, and lie on top of the Wall, where the Entablature ought to be raised. Also a kind of Terrass Walk, or even Floor on the top of a Building; from whence we may take a fair Prospect of the adjacent Gardens or Fields. So an Edifice is said to be cover'd with a *Platform*, when it hath no Arched Roof.

PLATFORM, in Fortification, is a Place prepared on the Ramparts for the raising of a Battery of Cannon; or it is the whole Piece of Fortification raised in a re-entring Angle. See *Battery*.

PLINTH *Palladio* calls the *Orle*, and *Blondell* the *Abacus*. The Word is also used for a like Member about the Capital of a Pillar, but then 'tis called always the *Plinth* of the *Capital*, and is placed just above the *Echinus* in the *Dorick*, and above the *Ovolo*, or quarter Round in the other Orders.

PLINTHUS or *Plinthis*, in Architecture, is taken for that Square Member which serves as Foundation to the Base of a Pillar: But *Vitruvius* calls the upper part, or *Abacus* of the *Tuscan* Pillar, a *Plinth*, because it resembles a square Tile.

Moreover, the same Denomination is sometimes attributed to a thick Wall, wherein there are two or three Rows of Bricks advanced in form of a *Plat-band*.

POLYGON *Exterior*, in Fortification, is the Distance of one Point of a Bastion from the Point of another, reckoned all round the Work.

POLYGON *Interior*, is the Distance between the Centres of the two Bastions.

PORTICO, or *Porch*, in Architecture, is a long Place cover'd either with a Vaulted Roof, or an even Floor, supported by Pillars. But this Word may be applied to signifie all manner of Diposition of Columns in a Gallery.

POSTERN, in Fortification, is a false Door usually made in the Angle of the *Flank*, and of the *Curtain*, or near the *Orillon* for private Sallies.

PROFIL (*French*) a Term in Painting, signifying properly a Face or Head set sideways, as usually on Medals, and such a Face is said to be in *Profil*, or in a Side View. 'Tis also spoken of the View of a Building; or City, *etc.* in Opposition to the Plan or Ground-plot of it; and so appears to have much the same sense as a Prospect of any Place, City, or Piece of Architecture, viewed sideways, and expressed according to the Rules of *Perspective*. This is called the *Profil* of such a Place, City, *etc.* Some call the Out-lines of any Figure its *Profil*, but that way of speaking is improper.

PROJECTURE, a Term in Architecture, signifying the Jutting or Leaning-out of any Part of a Building, the Coping of a Wall, *etc.* There the *Italians* call *Sporti*, and the *Greeks*, *Echphoras*; and in the General, all *Margents* which hang over beyond the *Scapus* of a Column, are called *Projectures*.

PRONAOS, or *Pronaus*, a Term used by Architects for a Church-Porch, or a *Portico* to a Palace, great Hall, or spacious Building.

PROPERTY, or *Propriety*, strictly speaking, is the highest Right that a Man hath or can have to any thing, and no ways depending upon any other Man's Courtesie: And this, none in our Kingdom can be said to have in any Lands or Tenements, but only the King in the Right of his Crown; because all the Land throughout the Realm is in the Nature of Fee, and held either mediately or immediately of the Crown.

This Word nevertheless is used for that Right in Lands and Tenements that Common Persons have, because it importeth as much as *utile Dominium*, tho' not *Directum*. And there are Three manners of Rights of *Property*, that is, Property *Absolute*, Property *Qualified*, and Property *Possessory*.

PSEUDODIPTERON, is an ancient form of a Temple, compassed about with but one row of Pillars, and which Row from the Wall, is at the distance usually of two Rows of Pillars.

PYCNOSTYLE, in Architecture, is a Building where the Pillars stand very close to one another, one diameter and a half of the Column being only allowed for the *Intercolumniation*.

R

RAKED *Table*, a term in Architecture. See *Table*.

RAMPART, in Fortification, is the mass of Earth which is raised about the Body of any Place, to cover it from Great Shot, and consists of several *Bastions* and *Curtains*; having its *Parapet*, *Platform*, *interior* and *exterior Talus*, and *Berme*, as also sometimes a Stone Wall, and then they say it is Lined. The Soldiers continually keep Guard here, and Pieces of Artillery are Planted for the Defence of the Place.

The Height of the *Rampart* must not exceed three Fathom, as being sufficient to cover the Houses from the Batteries of the Cannon: Neither ought its Thickness to be above ten or twelve, unless more Earth be taken out of the Ditch, than can be otherways bestowed.

The *Ramparts* of Half-Moons are the better for being low, that the small Fire of the Defendants may the better reach the Bottom of the Ditch; but yet it must be so high, as not to be commanded by the Covert-way.

RASANT *Line of Defence*, in Fortification, is a Line drawn from the Point of the Bastion along the Face, and prolonged till it come to the

Curtain, and therefore shows how much of the Curtain will clear or scour the Face. This is called also the *Second Flank*, the *Hanking*, or *Stringent Line*.

RAVELIN, in Fortification, is a small Triangular Work composed only of two Faces, which make a Saliant Angle, without any Flanks. It is generally raised before the *Curtains* or *Counterscarp*, and commonly called a *Half-Moon* by the Soldiers.

A *Ravelin* is like the Point of a Bastion with Flanks cut off. The reason of its being placed before a Curtain, is to cover the opposite Flanks of the two next Bastions. 'Tis used also to cover a Bridge or a Gate; and 'tis always placed without the Moat.

What the Engineers call a *Ravelin*, the Soldiers generally call a *Half-Moon*, which see.

REASON, in Mathematicks, the same *Proportion*, 'tis better called by the Latine Name *Ratio* to avoid confusing it with the common signification of the Word *Reason*.

REDENT, in Fortification, is a Work made in Form of the Teeth of a Saw, with *Saliant* and *Re-entring Angles*, to the end that one part may defend another. These sort of Works are usually erected on that side of a Place which looks towards a Marsh or River.

REDOUBT, in Fortification, is a small Fort of a square Figure, having no Defence but in the Front, its use being to maintain the *Lines of Circumvallation*, *Contravallation*, and *Approach*. In Marshy Grounds, these *Redoubts* are often made of Mason's Work for the Security of the Neighbourhood. Their *Face* consists of from ten to fifteen Fathom; the Ditch round about being from eight to nine Foot broad and deep, and their *Parapets* having the same thickness.

RELAIS, a *French* Term in Fortification, the same with *Berme*.

RELEIFE, or *Releivo*, is the Protuberant jetting or standing out of any Figures or Images above the Plain on which they are formed. And whatever Figures or Representaions are thus cut, stamped, or otherwise wrought, so that not the entire Body, but only part of it is raised above the Plain, is said to be done in *Releife*; and when the Work is low, flat, and but a little raised, 'tis called *Basse Releife*, or *Low Releife*. When a Coin or a Medal hath its Figure low and thin, and hardly distinguishable from the Plain, we say its *Reliefe* is low and weak; but when 'tis much raised, we say 'tis Bold, and its Releife is Strong.

RETIRADE, in Fortification, is a kind of Retrenchment made in the Body of a Bastion or other Works, which is to be disputed Inch by Inch, after the first Defences are Dismantled. It usually consists of two Faces, which make a Re-entering Angle.

RETRENCHMENT, in Fortification, is a Ditch bordered with its Parapet, and secur'd with Gabions or Bavins laden with Earth. It is sometimes taken for a simple *Retirade* in part of the *Rampart*, when the Enemy is so far advanced, that he is no longer to be Resisted, or beaten from the first Post.

REVERSED *Talon*. See *Talon.*

RIDEAU, in Fortification, is a Ditch, the Earth whereof is raised on its side, or a small elevation of Earth, extending it self in length on a Plain, which serves to cover a Post, being also very convenient for those that would besiege a Place at a near distance; and to secure the Workmen in their Approaches to the Fort of a Fortress.

ROMAN *Order of Architecture*, is the same with the *Composite*. 'Twas invented by the Romans in the Time of *Augustus*, and set above all the others, to shew (say some) that the *Romans* were Lords over all other Nations: 'Tis made up of the *Ionick*, and *Corinthian Orders*, and is more ornamental than either.

RONDEL, in Fortification, is a round Tower sometimes erected at the Foot of the Bastions.

ROOD, a Square Measure containing just a Quarter of an Acre of Land: Some confound this Measure with a *Rod*, which is the Length of 16½ Feet; and others with a Yard land, or the *Quatrona Terræ*, but both very erroneously.

A *Rood* is sometimes called a *Farthendele.*

S

To **SAIGNER** *a Moat*, is to drain the Water by Subterraneous Conveyances, to the end that Hurdles laden with Earth, or a Bridge of Bulrushes may be afterward laid upon the Mud that remains, and the Passage thereby consolidated.

SAPPE, in Fortification, formerly signified the undermining, or deep digging with Pick-axe and Shovel at the Foot of a Work to overthrow it without Gun-powder: Now it is used to signifie a deep Trench carried far into the Ground, and descending by Steps from top to bottom; so that it covers the Men sideways; and to save them from danger on the top, they use to lay a-cross it Madriers, that is thick Planks; or Clugs, that is, Branches of Trees close bound together, and then they throw Earth over all, to secure them from Fire. When a Covert-way is well defended by Musketiers, the Besiegers must make their way down into it by *Sappe*.

SAUCISSONS, are a sort of Faggots made of great Boughs of Trees bound together in the middle and at both ends. Their use is much the same with the Bavins or small Faggots, *viz.* to consolidate the Way for Carriages; to make Traverses or Parapets in Ditches full of Water, *etc.*

SCAMILLI *Impares*, in Architecture, are certain Blocks or *Zocco's* which serve to elevate the rest of the Members of any Column or Statue, which was placed before the Horizon, (*i.e.*) beneath the Projectures of the *Stylobata Cornices*, and, other *Scillies*; and will easily be conceived by considering the Pedestals of Statues, which do well represent them.

SCARP, in Fortification, is the Foot of the Rampart-Wall, or the sloaping of the Wall from the bottom of the Work, to the Cordon on the side of the Moat.

SCENOGRAPHY, in *Perspective*, the *Scenographick* appearance of any Figure, Body, or Building, is that Side that declines from, or makes Angles with that strait Line imagined to pass through the two outward convex Points of the Eyes, generally called by Workmen, the *Return* of a fore-right Side; and differs from the *Orthographick* Appearance in this, That the latter represents the Side of a Body or Building as it is seen, when the Plain of the Glass stands parallel to that Side: But *Scenography* represents it as it seems through a Glass, not parallel to that Side.

In Architecture and Fortification, *Scenography* is the manner of delineating the several parts of a Building or Fortress, as they are represented in Perspective.

SCIOGRAPHY, is the Art of *Shadows* or *Dyaling*: Also in Architecture, this Word is sometimes taken for the Draught of a Building cut in its Length or Breadth, to shew the Inside of it; as also the Thickness of the Walls, Vaults, Floors, Timber-works, *etc.*

SCOTIA, in Architecture, is a certain Member hollowed in form of a Demi-channel, which is placed between the *Torus* and the *Astragal* in the Bases of Pillars; as also sometimes under the *Larmier* or *Drip*, in the Cornice of the *Dorick* Order.

SCOTIA, a term in Architecture, the same with *Trochile*, which see.

SCROWLES, or *Volutes*, a term in Architecture. See *Voluta*.

SECTION *of a Building*, in Architecture, is understood of the Persil or Delineation of its Heights and Depths raised on the Plane; as if the Fabrick were cut asunder to discover the Inside.

SHOULDRING, in Fortification, is a Retrenchment opposed to the Enemies, or a Work cast up for a Defence on one side, whether it be made of heaps of Earth cast up, or of *Gabions* and *Facines*. A *shouldring* also is a square *Orillon* sometimes made in the Bastions on the Flank near the shoulder, to cover the Cannon of a Casemate. Again, it is taken for a Demi-bastion or Work consisting of one Face, and one Flank, which ends in a Point at the head of a *Horn-work*, or *Crown-work*: Neither is it to be understood only of a small Flank added to the sides of the *Horn-work*, to Defend them when they are too long, but also of the *Redents* which are raised on a strait Line.

SILLON, in Fortification, is an elevation of Earth, made in the middle of a Moat, to Fortifie it when too broad: It is otherwise called *Envelope*, which is the more common Name.

STAR, in Fortification, is a Work with several Faces, generally composed of from 5 to 8 Points, with Saliant, and Re-entring Angles Flanking one another; every one of its Sides containing from 12 to 25 Fathoms.

STEREOBATA, in Architecture, is the *Greeks* Word for the first beginning of the Wall of any Building, and immediately standing on the Foundation: This is wrongly confounded with *Stylobata*, which is the beginning of a Column, or its *Pedestal*.

STEREOGRAPHY, is the Art of drawing the forms of Solids upon a Plane.

STRUCTURE, is the combination or result of all those Qualities or Modifications of Matter in any Natural Body, which distinguish it from others, and are what, in other Words, is called the Peculiar *Form*, or *Texture* of it.

STYLOBATA, in Architecture, is the Pedestal of a Column or Pillar.

STYLOBATON, or *Stylobata*, in Architecture, is the same with the Pedestal of a Column. This is sometimes taken for the Trunk of a Pedestal, between the Cornice and the Base; and then called *Truncus*, as it also is by the Name of *Abacus*.

SWALLOW-*Tail*, in Fortification, is a *single Tenaille*, that is narrower towards the *Place* than towards the Country.

SYMMETRY, is usually taken in *Architecture*, for the Proportion required, according to the Rules of *Geometry*, to make all the Parts of any Structure compleatly agree to, and with the Whole.

SYSTYLE, in *Architecture*, is a Building where the Pillars stand thick, but not altogether so close as in the *Pychnostyle*; the Intercolumniation, or Distance between them, being only two Diameters of the Column.

T

TABLE, in Architecture, is a smooth and simple Part of a different Figure; but most commonly in form of a long Square, or of a Triangle.

Projecturing Table, is that which jetts out beyond the naked Face of a Wall, Pedestal, or any Part whereof it makes the Ornament: And a

Raked Table, is that which is hollow'd in the Square of a Pedestal, or elsewhere.

TÆNIA, in Architecture, is a Member of the *Dorick* Capital, which resembles the Shape of a square Fillet, and serves instead of a *Cymetium*, being fastned, as it were to a Capital below the *Triglyphs*, whereof it seems to be the Base.

TAILLOIR: See *Abacus*.

TALON, a small Member in Architecture, composed of a square Fillet, and a straight *Cymetium*. It differs from the *Astragal* which is a round Member, whereas the *Talon* consists of two Portions of a Circle, one without, and the other within; and when the Concave Part is uppermost, it is called *Reversed Talon*.

TALUS, or *Talut*, properly signifies any Thing that goes sloping, as the *Talus* of a Wall in Masonry, when its thickness is diminished by Degrees as it rises in height. But in *Fortification*, the *Talus* of a *Bastion* or *Rampart*, is the Slope allowed to such a Work whether it be of Earth or Stone, the better to support its Weight.

TALUS *Exterior*, of a Work, is its Steepness on the Sides of the Field; and is always made as little as possible, to prevent the Enemies *Scalando*, unless the Earth be bad, then it is absolutely necessary to allow a considerable *Talus* for its *Parapet*.

TALUS *Interior*, of a Work, is its Steepness on the Inside towards the Place.

TECHNICAL, is sometimes the same with *Artificial*, and expresses whatever relates to the Arts and Sciences, as the *Terms*, *Rules*, etc. So that the Terms of Art are commonly called *Technical Words*.

TEINT, is a French Term in Painting, signifying that Artificial Colour, of which any Object is composed.

TENAILLE, in Fortification, is a kind of Out-work resembling a *Horn-work*, but generally somewhat different, in regard that instead of two Demi-bastions, it bears only in Front a Re-entring Angle between the same Wings without Flanks; and the Sides are parallel: But when there is more bredth at the *Head* than at the *Gorge*, these *Tenailles* are called *Queuve d'yronde*.

TENAILLE *Double or Flank'd* is a Work whose Front consists of four Faces, making two *Re-entring Angles*, and three *Saliant*; the Wings or Sides of this Work being in the like manner correspondent to the Front of the *Gorge*.

TENAILLE *Simple*, is a Work having its Front form'd by two Faces, which make a Re-entring Angle, the Sides running directly parallel from the Head to the Gorge.

TENAILLE *of the Place*, is that which is comprehended between the Points of two neighbouring Bastions; that is to say, the *Curtain*; the two *Flanks* that are raised on the Curtain, and the two Sides of the *Bastions* which face one another: So that 'tis the same with what is otherwise called, *The Face of a Fortress*.

All Tenailles are defective in this respect, That they are not Flanked or Defended towards their *inward* or *dead Angle*; because the height of the Parapet hinders seeing down before the Angle, so that the Enemy can lodge himself there under Covert: Wherefore *Tenailles* are never made but when they want Time to make a *Horn-work*.

TENEMENT, signifies the House or Land that a Man holdeth of another, and when join'd with *Frank*, it contains Lands, Houses, and Offices, wherein we have Estate for term of Life, or in Fee.

TERRE-PLAIN, in Fortification, is the Plat-form or Horizontal Surface of the Rampart; lying level, only with a little sloap on the out-side for the Recoil of the Cannon.

It is terminated by the Parapet on that Side toward the Field, and by the *Inner Talus* on the other toward the Body of the Place.

TETRASTYLE, in Architecture, is a Building which hath four Columns in the Faces before and behind.

THEODOLITE, is an Instrument used in *Surveying*, and taking of Heights and Distances.

It consisteth of several Parts: As,

1. *A Circle of Brass*, divided into four Quadrants, representing the four Cardinal Points of the Compass, *East*, *West*, *North*, and *South*, and noted with the Letters, N, S, W, E.

Each of these *Quadrants* is divided into 90 Degrees, and Subdivided as the Largeness of the Instument will permit, commonly by *Diagonals*.

These four Quadrants are to be numbred by 10, 20, 30, *etc*. both ways beginning at the *North* and *South* Points, and ending with 90, at the *East* and *West* Points.

2. *A Box* and *Needle*, so conveniently contrived to stand upon the Center of the Circle, upon which *Center*, the *Instrument*, the *Index* with its *Sights*, must be made to turn about, and yet both the *Instrument*, and the *Box* and *Needle* remain firm. At the bottom of the *Box* there must a *Card* or *Mariners Compass* fixed, answerable to the Letters E. W. N. S. upon the *Instument*.

3. A *Socket* on the Back-side, (either Plain or with a Ball, which is much better) to be put upon the Head of a three-legged Staff.

4. A *Staff* to set the *Instrument* upon; the Neck at the Head whereof, must be made to go into the *Socket* on the Backside of the *Instrument*. Note: *This is an edited extract from a larger article*.

TONDINO, a Term in Architecture. See *Astagal*.

TORE, or *Torus* in Architecture, or as 'tis sometimes spelt *Thore*, is that round Ring which encompasses in the Column, between the *Plinth*, and the *List*. This is the third Member of the Base of a Column; it seems like a round Cushion, swelling out with the Weight of the Incumbent Pillar.

TORUS, in Architecture, is a round, thick Circle running round the Base of a Pillar, resembling the form of a great Ring.

TRABEATION, in Architecture, is the same with *Entablature*, *viz*. the Projecture on the Top of the Walls of Edifices which supports the Timber Work of the Roof. *Trabeation*, or *Entablature*, is different according to the several Orders, and comprehends the Architrave, Frize, and Cornice.

TRAVERSE, in Fortification, is a little *Trench* bordered with two *Parapets*, *viz*. One on the right Side, and the other on the left, which the Besiegers make quite thwart the *Moat* of the Place, to pass secure from *Flank*-shot, and to bring the Miners to the Bastions.

This *Traverse* is usually covered on top with *Hurdles* or *Bavins* laden with Earth, for Security from the Fire-Works, and differs from a *Coffer* only in this respect, That it is made by the *Besiegers*, and the other by the *Besieged*.

This Word is now often used for any Retrenchment or Line Fortifiy'd with Fascines, Gabions, Barrels, Bags of Earth, *etc*.

TRENCHES, in Fortification, are certain Moats or Ditches, which the Besiegers cut to approach more securely to the Place Attack'd, and are of several sorts, according to the different nature of the Soil; for if the adjacent Territory by Rocky, the Trench is only an Elevation of *Bavins*, *Gabions*, *Wool-packs*, or Shouldrings of Earth cast up round about the Place: But where the Ground may be easily open'd, the Trench is dug therein, and border'd with a Parapet on the Side of the Besieged. The Breadth of it ought to be from 8 to 10 Foot, and the Depth from 6 to 7.

These *Trenches* are to be carried on with winding Lines, in some manner parallel to the Works of the Fortress, so as not to be in view of the Enemy, nor to expose its Length to their Shot, which they call *Enfilading*; for then it will be in danger of being *Enfiladed*, or scoured by the Enemies Cannon: And this carrying of the Trenches obliquely, they call, carrying the Trenches by *Coudees* or *Traverses*.

They call it *Opening the Trenches*, when the Besiegers begin to Work upon the Line of Approaches; and this is usually done in the Night, and sometimes within Musket-shot, and sometimes within half or whole Cannon-shot of the Place, if there be no rising Ground about it, the Garrison Strong, and their Cannon well served. The Workmen that Open the Trenches, are always Supported by Bodies of Men against the Sallys of the Besieged; and sometimes those Bodies lie between them and the Place, as also on their Right and Left. The Pioneers sometimes Work on their Knees, and the Men that are to support them, lie flat on their Faces, in order to avoid the Enemies Shot: And the Pioneers are also usually covered with *Mantelets* or *Saucissons*. They also say, *Mount the Trenches*; that is, go upon Duty in them: And to *Relieve the Trenches*, is to Relieve such as have been upon Duty there. To *Carry on the Trenches*, is to advance them towards any Place.

TRIGLYPH, in Architecture, is a Member of the *Frize* of the *Dorick Order*, set directly over every Pillar, and in certain Spaces in the Inter-columniations. By their Triangular Gutters, (for they are always made with three *Glyphæ*, whence the Name;) they seem to have been deisgn'd to Convey the *Guttæ*, or Drops which hang a little under them: Some think they represent *Apollo's* Lyre.

TRINGLE, in Architecture, is a little Member fix'd exactly upon every *Triglyph*, under the Plat-band of the Architrave, from whence hang down the *Guttæ* or Pendant Drops in the *Dorick Order*.

TROCHILE, in Architecture, is that hollow Ring or Cavity which runs round a *Column* next to the *Tore*; vulgarly 'tis call'd the *Casement*: 'Tis frequently bordered or rather shut in with *Lists*. 'Tis often called *Scotia*, from its shady dark Appearance.

TROPHY, in Architecture, is an Ornament which represents the Trunk of a Tree charged or encompassed all round about with *Arms* or *Military Weapons*, both *Offensive* and *Defensive*.

TUSCAN *Order* of Architecture; so called, because Invented in *Tuscany*, an eminent Part of *Italy*. 'Tis the most plain of all the 5, and is seldom used but in some Country Building, where there is no need of any *Order* but one; or else in some very great Building, such as an *Amphitheatre*, or the like, where all the Orders are designed to be used.

The Columns here, together with Base and Capital, are to be 7 Modules in Length, and to have their Thickness diminish'd by degrees to a quarter part. When these Columns stand alone, the space between them or the *Inter-columns* ought to be very large.

Their *Pedestals* ought to be one Module high; and the *Base* of the *Column* ought to be of the Height of half its Thickness; and that Height divided into two equal parts, makes the *Orle* or *Plinth*, which is made with a kind of Compass; the other is divided into 4 parts, of which one is for the *List*, the other for the *Tore*, or thick round Circle of the Pillar.

The *Base* hath a *Saille* or Bearing out, which is the 16th part of the diameter of the Column.

The *Capital* is in Height half the thickness of the Column at the Base; and is divided into three equal parts, (according to *Ozanam*) of which, one is for the Abacus, the other for the *Oeuf*, as they call it in French, *i.e.* the *Ornament*; and the third part being divided into 7 Divisions, the first makes the *List* under the *Ornament*; and the other six serve for the *Gorgerin Collier*, or *Frize*. The *Astrogal* is the height of the *List* below the *Ornament*.

TUSCAN *Work*, in Architecture, is the most simple and rude of the Five ancient Orders of Pillars. See *Column* and *Order*.

TYMPAN, in Architecture, is that part of the bottom of the *Frontons*, which is enclosed between the *Cornices*, and answers the Naked of the *Frize*.

TYMPAN *of an Arch*, is a triangular Table placed in its Corners.

The most simple of these *Tympans*, have only a Table hollow'd, sometimes with the branches of Lawrel, Olive-tree, or Oak, or with Trophies, and are conformable to the *Dorick* and *Ionick* Orders. But the richest, are adorn'd with flying Figures, as *Fame*, or sitting Figures, as those of the *Cardinal Virtues*, and are proper for the *Corinthian* and *Composit* Orders.

Tympan, is also attributed to the Panels of Doors in Joyners Work, and to the *Dye* or Square of Pedestals.

U

UNION, (a Term among Painters) is the mutual Agreeableness and Sympathy of the Colours in a Piece of Painting.

V

VESSELS, in Architecture, are certain Ornaments, usually set over the *Cornices*, and so named, because they represent divers sorts of *Vessels*, which were in use among the Ancients.

VISUAL *Point* in a Perspective, is a Point in the Horizontal Line, wherein all the Ocular Rays unite; and in which, and all the others that are parallel thereto. As if a Man stood in a long strait Gallery, wherein looking directly forward, the Sides, Floor, and Ceiling at last seem to be united, and to touch one another in a Point or common Centre.

VOLUTA, in Architecture, is that part of the Capitals of the *Ionick*, *Corinthian*, and *Composite* Orders, which is supposed to represent the Bark of Trees twisted, and turned into Spiral Lines; or, as some say, the Head-dresses of Virgins in their long Hair. *Voluta's* are different in these three *Orders*: Those that appear above the Stems in the *Corinthian* Order (according to *Vitruvius*) are 16 in Number in every Capital, whereas there are only 4 in the *Ionick* Order, and 8 in the *Composit*. But these *Voluta's* are more especially remarkable in the *Ionick* Capital, representing a kind of Pillow or Cushion laid between the *Abacus* and the *Echinus*, as if it were to be fear'd lest the Weight of the *Abacus*, or the *Entablature* above it, might break or deface the *Echinus*; whence the same Ancient Architect took occasion to call the *Voluta*, *Pulvinus*, or Bolster.

W

WAY *of the Rounds* in Fortification, is a Space left for the Passage of the *Rounds* between the Rampart and the Wall of a Fortify'd Town. But it is not so much in use, because not having a Parapet above a Foot thick, it may be soon overthrown by the Enemies Cannon.

WINGS, in Fortification, are the large Sides of Horn-works, Crown-works, Tenailles, and the like Out-works; that is to say, the Ramparts and Parapets, with which they are bounded on the Right and Left, from their Gorge to their Front. These Wings or Sides are capable of being Flank'd either with the Body of the Place, if they stand not too far distant; or with certain Redouts, or with a Traverse made in their Ditch.

Z

ZOCCO: See *Plinthus*.

ZOCLE, is a Square Member in Architecture, being lower than its Breadth, which serves to support a Pillar, or any other part of a Building, instead of a Base, Plinth, or Pedestal.

Continued Zocle, is a kind of continued Pedestal, on which a Structure is raised, but hath no Base, or Cornish.

7 *Lexicon Technicum* Volume II

Title Page

Dedication

The Introduction

Lexicon Technicum:

Or, An UNIVERSAL

English Dictionary

OF

ARTS and SCIENCES

EXPLAINING

Not only the TERMS of ART, but the ARTS Themselves

VOL. II.

BY

JOHN HARRIS, D. D. Secretary to the *Royal-Society*, and Chaplain to the *Lord High-Chancellor* of GREAT-BRITAIN.

LONDON:

Printed for *Dan. Brown*, *Tim. Goodwin*, *J. Walthoe*, *Joh. Nicholson*, *Benj. Tooke*, *Dan. Midwinter*, M. *Atkins*, and *T. Ward*.

MDCCX.

TO THE

RIGHT HONOURABLE

WILLIAM Lord *COWPER*,

Baron of *WINGHAM*,

AND

Lord High-Chancellor

OF

GREAT-BRITAIN.

MY LORD,

T*HE great Honour and Advantage which the former Volume of this Work receiv'd from the Patronages of His Royal Highness, the Late Prince* George *of* Denmark, *encourages me to Dedicate This to Your Lordship; as I am also oblig'd in Duty and Gratitude to do, for the many Favours I have received from You; and especially for the great Honour and Happiness of having been so long known, and so near to You: For this, my Lord, will save me from the Fate of those Dedicators, who attempt Characters above Panegyrick: The Publick Part of Your Lordship's all the World knows and* *admires; and tho' I see every Day That which renders Your Lordship the* Instruction*, as well as the* Delight *of all that have the Honour of Your Conversation, yet I* can *no more describe it, than I* dare *attempt it: I shall therefore, my Lord, entirely forbear; fearing as much doing Violence to Your Modesty, as Injustice to Your Merit.*

I am,

My LORD,

Your Lordship's

Most Dutiful, and

Most Humble Servant,

JOHN HARRIS.

The Introduction.

The Design of an *Introduction* to a Book, being to lead the Reader into it, and to acquaint him what it contains; I shall observe that Method here.

He is therefore to understand, that when I first began this Work, I foresaw the *Design* could not be accomplish'd in One Volume; but however, I thought it more proper to Publish one Volume first, than to deferr Printing *any thing* at all of it till the Whole was finished: For as it is easie to see, that new Matter will continually ocurr in a Design of this Nature, and, consequently, that there can be no such thing as a *Perfect Book* of this Kind; so I thought it better to send out an *Imperfect one* than none, and to afford *some* Help to Mens Improvement in Philosophy and Mathematical Studies, rather than leave them to stay four or five Years for a *more Compleat* one. And the unprejudic'd Part of the Learned World have been so kind, as to take my Endeavours as I meant them; and by taking off almost Two Impressions of the former Volume, and numerously Subscribing to this, make me hope, the Pains I have taken, and the Time I have employed this way, may be some Use and Benefit to Mankind, and to the Improvement of Solid and Substantial Philosophy.

In this Second Volume, as I promised both in the *Preface* to the First, and also in the *Proposals* for This, the *Matter* is entirely *New*, and without any Repetition, that I know of, of any thing in the *Former*; and that shall be my Method, if ever I publish any thing further in this Way.

The Reader will find here many Parts of Natural Philosophy and Anatomy largely treated of, which were either but just nam'd, as it were, or entirely omitted in the former: As in Particular, the Affair of *Animal Secretion*; into which Dr. *Keil's* Book on that Subject, Publish'd since my Account was Printed, will let you yet further; Discoursees on Thunder, a Vacuum, Vapours, Water; with a large Account of Sound, Echoes: The Transmutation of Bodies into one another; the Nature of Light and Colours, the Rays of Light, the Double Refraction found in the Island *Chrystall*; Elasticity, Electricity, the Cohesion of the Parts of Bodies, *etc.*

And in Natural History, I have here given Schemes of Birds, Fishes, Insects, Quadrupeds, Roots of Plants, *etc.* by which they are ranged and distributed into their proper Orders.

Here are also inserted pretty large Accounts of the Ear, Stomach, Spleen, Skin, *Vena Porta*, Lungs, Gall, and Respiraton, and of the New Theory of Generation; with Descriptions of the Lymphæducts and Glands, and two very fine and large Copper Plates of the Veins and Arteries of a Human Body, which were drawn from the Original Tables presented by that great Promoter of Useful Learning, *John Evelyn* of *Deptford*, Esq; to the *Royal Society* of *London.*

You have here also a further Account of that most amazing Property, the *Attraction* of the Particles of Matter one towards another, first discover'd by that Incomparable Mathematician and Philosopher Sir *Isaac Newton*; who by the wonderful Discoveries He hath made about the Nature of *Light* and *Colours*, hath open'd a New World in Natural Philosophy, as by His Method of Fluxions, *etc.* he had before done in Mathematicks, Mechanicks and Astronomy; and hath sufficiently shewn, that what he said in his *Preface* to his *Principia Philos. Math.* in these Words; *Multa me movent ut non nihil suspicer cætera Naturæ Phænomena, ex viribus quibusdam pendere posse, quibus Corporum particulæ, per Causas nondum cognitas, vel in se mutuo impelluntur, et secundum Figuras regulares cohærent, vel ab invicem fugantur et recedunt*, was what he very well knew *then*, tho' express'd with that Caution and Modesty, as is so peculiar to that Excellent Man.

This was Printed in the Year 1687, and the *Queries* at the End of his *Opticks*, and especially as since enlarged by him in the *Latin* Translation of it, do sufficiently shew his Thoughts to have been long ago employ'd on this most useful Subject; and from whence those Proportions took their Rise, which those Ingenious and Industrious Brothers, the *Keils*, have Publish'd about this Affair of Attraction: But however, to do the Illustrious Author yet further Justice, I have, with his Leave, at the End of this *Introduction*, Printed a *Latin* Paper of his *De Acido*, with a Translation of my own; and which, tho' never Publish'd before, was given by him to a Friend, as long since as the Year 1692, and which I wish had come to my Hands sooner, to have been inserted in this *Lexicon*, under the proper Head, *Acids*.

And give me leave here further to inform the Reader, That there is now Printing a *Latin* Mathematical Treatise or two, of Sir *Isaac Newton's*, which were written many Years ago, and which by their Date, will sufficiently determine, whether the *New Methods of Fluxions* were known first to him, or Mr. *G. Leibnitz*. But to go on; In this Second Part, I have been very full and particular in Astronomy; Having not only from Mr. Hayes's Excellent Book of Fluxions, given a short System of the New Astronomy, but have also from the Ingenious and Learned Mr. *Halley*, *Savilian* Professor of Geometry in the University of *Oxford*, inserted a *Synopsis* of the Doctrine of Comets; and from the Prelections of Mr. *Whiston*, Mathematic Professor of the University of *Cambridge*, such Astronomical Tables as are necessary for Calculation, illustrating largely the Uses of them, with proper Examples, in the most useful Astronomical Problems, and in the Calculation of Solar and Lunar Eclipses, and in those of the Satellites of *Jupiter*, *etc.*

And because I would have these Two Volumes to serve as a kind of small Mathematical Library, and prevent in some measure the Necessity and Charge of buying many Books on these Subjects; I have also, as I design'd at first, given you in this Volume very good Tables of Logarithms, Sines, Tangents and Secants, with a full Account of the Nature, Use and Application of them; so that nothing will be wanting here to compleat Trigonometry, both Plain and Spherical, and the Practice of it in Navigation, Dialling and Astronomy, *etc.*

I have here also given a full and clear Account of the Nature, Construction and Uses of all the Lines which are usually drawn on any Mathematical Instruments, Rules, Scales, *etc.* and how to use those Instruments, on which they are drawn, in the several Parts of Practical Mathematicks for which they were design'd.

And I have here and there inserted such useful Tables for the Calculation of Interest, Annuities, Purchases, Reversions, *etc.* as I found most easily and readily subservient to those Uses. And because of the very many and excellent Uses of the Table of Incomposite Numbers, and its being Printed no where but in *Brancker's Algebra*, an obscure Book, and out of Print; I have here given it you entire, and, I believe, correct; being desirous to prevent so useful a Table from being lost or buried in Obscurity.

Here is also an Account of the Rise, Invention and Progress of the Art of Printing; with a Description of the Tools and Instruments subservient thereunto.

I have added also a further Account of the Phænomena of Prisms and the Rain-bow; with a short System of Opticks from Sir *Isaac Newton*; and many Improvements in Microscopes and Telescopes; the Art of Perspective, and the Methods of Projective, Reflective and Refractive Dialling.

In Mechanicks, besides the Account of the Five Powers and Demonstrations of the Nature and Principles of that Science, here are many things added about Centres of Gravity and Oscillation; the Resistance of Mediums; with the Description and Uses of all such Mechanical Instruments. Tools and Engines as are used in Architecture, Fortification, Gunnery, or any Mathematical Arts and Sciences.

In Geometry, here are great additional Improvements; as two Treatises of the *Order* and *Quadrature* of Curves, written by the Incomparable Sir *Isaac Newton*; a Treatise of Conick Sections, translated from the Posthumous Book of the Marquis *de L'Hospitall*, with many other things relating to the Properties of Curves, scatter'd up and down under particular Words; such as *Cycloid*, *Helicoid*, *Retrogression of Curves*, *Transcendental Quantities*, *etc.* Here are also added many new things in Arithmetick and Algebra, and in the Doctrine of Fluxions; as about *Alternations*, *Combinations*, the *Laws of Chance* in Play, *Infinite Series*, and *Political Arithmetick*; and about the *Roots of Equations*, *Renewing of Leases*, *Reversions*, *etc.*

I have also here given a large Account of the Ways of Finding, Dressing, Melting, *etc.* of all the several Ores from whence our Metals are taken; describing also the Works, Engines, Tools and Terms of Art used by Miners; as also the Ways of making Salt, Allum, Coperas, Vitriol and such like Mineral Productions.

In Musick, here is given an Account of the Nature and Grounds of Harmony, from Dr. *Helder*, and a new short System of Musick by the Ingenious Mr. *Perk.*

In Navigation I have added many things; as the Way of finding and allowing for the setting of Currents; an Account of the Power of the Winds on the Sails of Ships, and of the Signals used at Sea, both by Day and Night, in Sailing, and in an Engagement; together with a new Traverse Table, and its Use and Application: And at the End you will find two very accurate Cutts of the *Inside* and of the *Rigging* of a First Rate Man of War; with the several Parts describ'd, and referr'd to by proper Letters and Numbers.

I have also inserted Tables of the Sun's Place, Right Ascension, and Declination; together with a Catalogue of the Right Ascension, Declination, Longitude, and Latitude of about fifty of the Principal Stars; which Mr. *Hodgson* supplied me with; as also a Table of the Longitude and Latitude of the most eminent Places on the Globe; all which will be very useful in Astonomy, Navigation and Dialling.

I have added also a Copper-Plate, describing a new *Hydrostical-Ballance*, which is very ready and expeditious, to find the Specifick Gravity of Bodies, as is there shewn.

I have here, under the Word *Microscope*, given the Reader the Figure, Nature, Use and Advantages of Mr. *Willson's* Glasses of that kind, which I could only just mention in the *Introduction* to my former Volume, and which I think my self oblig'd again to say, are the most ready, commodious, and universaly useful, of any Microscopes, I ever saw.

And that I might do further Justice to our Excellent Mathematical Instrument-Maker Mr. *John Rowley*, in *Johnson's-Court* in *Fleet-Street*, I have given the Reader a Plate of a New Sextant lately made by his most accurate Hand, for the Observatory now building in *Trinity* College in *Cambridge*; and which for its universal Use, far exceeds any Astronomical Instrument ever yet made; as you will easily perceive by the Description of it, which I have added to the Figure.

I have only one thing more to acquaint the Reader, and that is, That the Ingenious and Accurate Mr. *Derham*, Rector of *Upminster* in *Kent*, and Fellow of the Royal Society, hath lately obtain'd from *Florence* an Account of the Measures and Weights used there: And we find by Mensuration and Trial, that the *Florentine Semibroccio*, or Half-Brace, is in Length 11.475 of our *Inches*, or in Foot-Measure .956 of our *Foot*: And the *Florentine* Ounce is 17 *Peny-weight*, 12 *Grains*, and three Fourths of a *Grain Troy-Weight*.

8 *Lexicon Technicum* Volume II

Architectural subjects portrayed in the letterpress articles

Our inventory of letterpress articles in the series *The Encyclopaedic-Dictionary in the Eighteenth Century*: *Architecture, Arts and Crafts* is composed of subjects derived from various branches of architecture and allied arts and crafts. The following short list of headwords is provided here to give the reader an outline impression of the range of material that is included in the letterpress articles:

Architectural terms, arch construction, building construction, building materials, cabinet work, carpentry, classical orders, decorative art, design principles including aesthetics, domes, drainage, fortification, furnishings, furniture, gardening, joinery, landscape terms, military art, mouldings, ornament, painting, perspective, sanitation, sculpture, surveying, terminology of the building trades and crafts, tools and equipment used in building construction, and vaults.

LEXICON TECHNICUM:

OR, AN

UNIVERSAL *ENGLISH*

DICTIONARY

OF

Arts and Sciences.

VOL. II.

A

ABACUS: Besides the former Account of the *Abacus*, I find that *Mauclerc*, in the Ionick Order, designs what the Work-men call an O.G., with a Fillet over it for the *Abacus*: Which Fillet is half the breadth of the O.G., and he calls it the *Fillet of the Abacus*. In the *Corinthian* Order, he Describes the *Abacus* as a seventh part of the whole Capital. *Palladio* calls the *Plinth* about the *Boultin* (or *Eckinus*) the *Abacus*: Which from its Form, saith he, is commonly called *Dado* or the Dye; and this is $\frac{1}{3}$ of the whole height of the Capital.

Scamozzi calls a certain Hollow or Casement, which is the Capital of the Pedestal of the *Tuscan* Order, by this Name *Abacus*.

ABREUVOIRS, are the Interval or Spaces between the Stones, or the Joints where the Mortar lies in any Pillar, Wall, or Building.

ADZ, is an Instrument whose Blade is made thin and something bending, and hath not its Edge Parallel to its Handle, as the Ax and Hatchet hath, but placed athwart to it. It is Ground to a Basil on the inside to its outer Edge. Its general Use is to take off thin Chips of Timber or Boards, and such Irregularities as the Ax can't come at, and where a Plane, tho' *Rank Sett*, will not make riddance enough. 'Tis much us'd in taking off the Irregularities of Floors when they are pinned and framed together; and sometimes on Posts framed upright, and ranged in with other Work framed to them, where the Edge of the Ax can't come.

ANCHORS, in *Architecture*, is a certain kind of Carving, in the Form of an Anchor or Arrow Head, which is placed by way of Ornament to the *Boultins* of Capitals of the *Tuscan*, *Dorick* and *Ionick* Orders; and also to the *Boultins* of Bed Mouldings, of the *Dorick*, *Ionick* and *Corinthian* Cornishes. The Anchors and Eggs being placed Alternately.

ANNULETS in Architecture; this signifies a narrow flat Moulding which is common to other Parts of a Column, the Bases, *etc*. as well as the Capital. And 'tis the same Member which sometimes is called a *Fillet*, a *Listelle*, a *Ceincture*, a *Supercitium*, *Liste*, *Tince*, *Square*, and *Rabit*.

ANTÆ, the same with *Antes*. See Vol. I.

ANTIPÆGMENTS, in Architecture, are the Ornaments or Garnishings in Carved Work which are set on Architraves, whether of Wood or Stone.

ANTIQUE, is a Word much used by Architects, Sculptors, and Painters, and by it they mean all the Ancient Pieces of Architecture, Sculpture, and Painting from the Time of *Alexander the Great* to the Irruption of the *Goths*; as also *Intaglias* within that Time: All which they call *Antique*; and whatever is done in Imitation of the Great Masters of that Age, they say is after the *Antique Manner*.

APERTIONS, in Architecture, are the Openings in any Buildings; such as Doors, Windows, Stair-cases, Chimneys, Outlets, or Inlets for Light, Smoak, *etc*.

ARCHES, are part of the inward support of any Superstructure, and they are either Circular, Ecliptical or Streight. Of the Circular Arches of Bridges, *etc*. some are such as the Work-men call *Skeen* or *Scheme*, which are flatter Arches, less than a Semi-circle. Some Circular Arches are such as those in our *Gothick* Buildings, *di Tarzo and di quarto acuto*, as the *Italians* call them, or as we say of the third and fourth *Point*. Because they consist of two Arches of a Circle (meeting in an Angle at the Top) and drawn from the division of a Chord into three and four, or more Parts at pleasure. *Elliptical* Arches were formerly much us'd instead of Mantle Trees in Chimneys: They had a *Key-stone*, and *Chaptrels* or *Imposts*, and consisted of two Hanses and a Scheme. *Strait Arches* us'd over Windows and Doors, *etc*. having plain strait Edges both upper and under; these Edges are parallel, but both the Ends and Joints do all point towards a certain Center. They are now usually about a Brick and a half thick, which when rubbed is about 12 Inches. The Work-men call the Levelling End of this Arch the *Skew-back*; and the level Joints between the Courses of Bricks in the Arch, they call the *Sommering*.

ARCHITECTURE. Some Writers on this subject.

Vitruvius in *Latin* at *Amsterdam* 1649. *Fol.*

In *English* by Mr. *Christopher Wase*.

In *French* by Mr.*Claude Perrault*, *Paris* 1673. *Fol.*

Course d'Architecture Enseigni dans l'Academie Royal d'Architecture. Par Mr. Fran Blondell, à Paris 1675. *Fol.*

Mr. *Evelyn's* Parallel of *Architecture*, last Edit. 1706. *Lond. Fol.*

Adami Boecleri Architectura curiosa nova cum multis Figuris. Norimberg.

Albert Dureri, *Architectura & Geometria. Paris* 1535.

Potsis Architecture in 2 *Vol. Fol.* in *Latin* and *Italian*, and lately done into *English.*

L'Architecture par Jacques Androuet du Cercendi Paris 1615.

L'Architectura d'Andria Palladio Venet. 1642. *con Fig*. This is Translated into *English*.

Vitruvius and *Vignola*, abridged by Mr. *Perault.*

Palladios's Architecture.

Scamozzi's } Architecture.
Wotton's }

ASHLAR, is the Work-mens Term for Common or Free-stone, as it comes out of the Quarry.

ASPECT *Double*, is a Term in Painting, us'd for a single Figure representing two or more different Objects; which is caused either by a change of the Position of the Eye, or by Angular Glasses, *etc.*

ATTITUDES, in Painting or Sculpture, are the proper Postures that the Figure should be placed in, so as agreeably to answer the design of the piece.

B

BACK *and Bottom Nails*, are such as are made with flat Shanks that they may hold fast, and yet not open the Grain of the Wood. They are used for nailing Boards together for Coolers, for Gutters to save Water under the Eaves of a House, or for any Vessel made of Planks or Boards, and design'd to hold Liquors.

BALDACHIN, in Architecture, is a Building in form of a Canopy or Crown supported by Pillars, often serving for the Covering of an Altar. Some also call the Shell over a Door by this Name, and pronounce it *Baldaquin*.

BALLISTERS, are a Row of small Pillars which support a Rail. They are sometimes called *Bannisters*, and are placed on Stairs, and in the Front of Galleries, Balconies, *etc*.

BANK, is the Carpenter's Term for a Piece of Firwood unslit from 4 to 10 Inches Square, and of any Length.

BARBICAN, is a Term in Architecture taken from the *French*, and signifies any Outwork belonging to a great Building.

BASIL is the sloping Edge of a *Chissel*, or of the Iron of a Plane used in Joinery.

BATTEN is the Workmens Name for a scantling of Wooden Stuff from two to four Inches broad, and about an Inch thick; the Length is pretty considerable, but undetermined as to Numbers.

BEAD, in Architecture, is a Moulding which in the *Corinthian* and *Roman* Orders is cut and carved into short Embossements, which look like Beads born in Necklaces: And sometimes an *Astragal* is thus carved.

A *Bead plain* is sometimes set also on the Edge of each *Fascia* of an *Architrave*. Its Convexity is usually about a Quarter of a Circle, and differs from a *Boultine* only in not being so large. A Bead is often placed on the lining Board of a Door-Case, and on the upper Edges of skirting Boards.

BEAM, in any Buiding, is a Piece of Timber lying across it, and into which the Feet of the principal Rafters are framed. No Building hath less than two of these Beams, *viz*. one at each Head; and into these Beams the Girders of the Garret Floor are framed; and if it be a Timber Building, into them the *Teazle Tennons* of the Posts are also framed. These *Teazle Tennons* stand at Right-angles to those which are made on the Posts to go into the *Raisons*, and the *Relish* or Cheats of these Teazle Tennons stand up within an 1½ Inch of the Top of the Raison; and then the Beam is *cauked* down (which is all one as Dove-tailing across) till the Cheeks of the Mortices in the Beam conjoin with those of the Teazle Tennon on the Posts.

BEARER, in Architecture, is a Post or brick Wall, which is trimmed to *shorten its bearing*, or to prevent its bearing with the whole Weight at the Ends only.

BEDMOULDING, in Architecture, is a Term used by the Workmen for those Members in a Cornish which are placed below the coronet or Crown; and now-adays a Bedmoulding usually consists of these 4 Members, 1. an O.G., 2. a List, 3. a large Boultine, and 4thly, under the Coronet another List.

BINDING *Joysts*, in Architecture, are those Joysts in any Floor into which the *Trimmers* of the Stair-Cases and Chimney-Waies are framed; and these should be stronger than common Joysts.

BOULTINE in Architecture is the Work-mens Term for a convex Moulding, whose Convexity is just a ¼ of a Circle: 'Tis plac'd next below the *Plinth*, in the *Tuscan* and *Doric* Capital.

BRACE in Architecture, is a piece of Timber framed in with Bevil Joynts, and used to keep the Buiding from swerving either Way: When a Brace is framed into the Kinglesses and principal Rafters, 'tis called by some a *Strut*.

BRADS, are Nails us'd in Building, having no spreading Heads as other Nails have: Of these some are called *Joyners Brads*, and are for hard Wainscot; others *Batten Brads*, and are for soft Wainscot; and some *Bill Brads* or *Quarter-heads*, which are us'd when a Floor is laid in hast, or for shallow Joists which are subject to warp.

BREST is a Term in Architecture used by some for that Member or Column which is called the *Thorus* or *Tore*.

BREST *Summers*, in a Timber Building, are the Pieces in the outward Parts of any Building, and in the middle Floors, (not in the Garret, nor in the Ground Floor) into which the *Girders* are framed.

BRICKS. The several kinds of Bricks used in Architecture are these: 1. *Compass Bricks*, which are of a Circular Form, and are used in steyning of Wells. 2. *Concave* or hollow *Bricks*, being on one Side flat like a common Brick, but on the other hollowed: They are used to convey

Water. 3. *Cogging Bricks* are used to make the indented Work under the Coping of Walls built of great Bricks. 4. *Coping Bricks*, which are formed on purpose for coping of Walls. 5. *Dutch* or *Flemish Bricks*, used to pave Yards and Stables, and for Soap-Boylers Fatts, and for Cisterns. 6. *Clinkers* are such Bricks which are *glazed* by the Heat of the Fire in the making. 7. *Feather-edg'd Bricks* are like the common Statute Bricks, only they are thinner on one Edge than on the other, and are used to pen up the Brick Pannels in Timber Buildings. 8. *Didoron*, was a Brick used by the Ancients of 1½ Foot, or 2 spans long, (whence the Name) and one Foot broad. This was the smallest sort of Brick used by the *Greeks* in their private Houses, for there was a larger sort in use in their publick Edifices, which they called, 9. *Pentadoron*, which was 3 Foot 9 Inches long, and 1 Foot broad. 10. *Samel* or *Sandal Bricks*, are such as lie outmost in a Kiln or Clamp, and consequently are soft and useless as not being thouroughly burnt. 11. *Great Bricks* are 12 Inches long, 6 broad, and 3 thick. The Weight of one about 15 Pound; so that 100 will weigh 1500, and 1000 of them 15000 Pound. Their Use is to build Fence Walls, together with, 12. *Pilaster* or *Buttress Bricks*, which are of the same Dimensions with them, only they have a Notch at one End of half the Breadth of the Brick. Their Use is to bind the Work at the Pilasters of Fence Walls which are built of great Bricks. 13. *Paving Bricks* or *Tiles*, these are of several Sizes in several Counties and Places. 14. *Place Bricks* are such as are made in a Place made on purpose for them near the Building they are to be used in. 15. *Statute* or *small common Bricks*: These ought to be 9 Inches long, 4½ broad, and 2½ thick; 100 of these usually weighs about 550 Pound, and consequently 1000, 5500 Pound, and about 407 in Number are a Tun Weight. These are commonly used in paving of Cellars, Hearths, Sinks, *etc.* 30 or 32, if true Measure, will pave a Yard Square, and 330 will pave a Square of 100 Foot, laid flat; but if laid edge-wise, they must be double in Number.

Bricks are burnt either in a *Kiln* or a *Clamp*: Those that are burnt in a *Kiln*, are first set or placed in it, and then the Kiln being covered with Pieces of Bricks, they first put in some great or cord Wood to dry the Ware with a gentle Fire, which is continued till the Ware is pretty dry; which they know by the Colour of the Smoaks turning from a whitish Dark to a black transparent Smook; they then put in no more Wood, but proceed to *burn the Bricks* with Bush, Furze, Straw, Heath, Brake or Fern Faggots, having first damm'd up the Mouth of the Kiln with their *Shinlog* as they call it (which is Pieces of Bricks piled up one on another, and then closed up with wet Brick Earth instead of Mortar) only leaving just Room to put in a Faggot: They then continue to put in more and more Faggots, till they make the Kiln and its Arches look white with Heat, and that the Fire begins to appear at the Top of the Kiln. Then they begin to slacken the Fire for about half an Hour or an Hour, and so let all cool by Degrees. The Ware will be burnt usually in about 48 Hours. But now-a-days about *London*, they usually burn Bricks in *Clamps*, which are built of the Bricks to be burnt, something after the manner of the Arches in Kilns, *viz*. with a Vacancy between each Brick's Breadth, *etc*. for the Fire to play through; but with this Difference, that instead of arching, they *truss* or *span* it over, by making the Bricks project one over another on both Sides the Place for the Wood and Coals to lie in, till they meet, and are bonded by the Bricks at the Top which closes all up. The Place for the Fuel is carried up strait on both Sides till about 3 Foot high, then they fill it almost with Wood, and over that lay a Covering of Sea-coal; and then they over-span the Arch: But they strew sea-coal also at the Clamp, betwixt all the Rows of Bricks; then they fire the Wood and then the Coal, and when all is burnt out, they conclude the Bricks burnt enough. *Builder's Dictionary*.

BRIDGE *of Communication*, is a Bridge made over a River, by which two Armies or Forts which are separated by that River, have a free Communication one with another.

BUILDING. There are some good Rules and Directions about Building which are given in *Moxon's Mechanick Exercises*, and other Books of that kind, which 'tis very proper every Gentleman shou'd be acquainted with, that he may not be cheated or imposed upon by his Workmen: And those are such as these; First, with respect to the Foundation, Care ought to be taken, that after the Cellars are dug, if there be any, or if none, after the Trenches are dug in which the Walls are to stand, to try all the Foundations either by a Crow, a Rammer, or which is best, with a *Borer*, such as Well Diggers, Miners, *etc*. use, whether they are thoroughly sound, and are fit to bear the Weight that is to be laid upon them. If the Foundation be not very loose, it may be cured by ramming in great Stones with an heavy Rammer, the Stones being placed close together, about a Foot wider on each Side of the Trench than the Breadth of the Wall is to be; but if it be so loose that this will not do, you must get good Pieces of Oak, whose Length must be the Breadth of the Trench, or about 2 Foot longer than the Breadth of the Wall, and these should be laid across the Foundation, about a Foot asunder, and being well ramm'd down, lay long Planks upon them, which must be pinn'd down to them: And if this be not sufficient, Piles of a much greater Length must be driven down so deep till they reach good Ground, and then strong Planks pinn'd down to their Tops *etc*. If the Foundation be faulty only in Patches, there may Arches be turned over the insufficient Places, which will take off the Weight from bearing upon them. And sometimes it may be proper to use *inverted Arches*.

The Foundation being well secured, the next Care is to see that the Walls be all made of the same Thickness as they are in the Design; for which Purpose, there should be Numbers to express their Thickness in the Draft.

The Walls also must be well wrought and *bounded* as they call it: In order to which the Mortar must be well made of good well burnt Lime and sharp Sand; the usual Proportion, is a Load of Sand of 36 Bushels to a Hundred of Lime of 25 Bushels, or a Bushel of Lime to one and an half of Sand; but when the Sand is not very sharp, there must be a greater Proportion of it to the Lime.

In slacking of Lime Care shou'd be taken, that though it be wetted every-where a little, yet it be not over-wetted; and every Layer or Bed of Lime should be covered with Sand to keep in the Steam, that it may not fly away, but be forced to mix itself with the Sand; and this will make the Mortar much stronger. The Mortar also must be well *beaten* with a Beater three or four times over before 'tis used, that the Knots of the Lime may be all broken, and the Lime and Sand mingled very well together: And it will be still better, if after the first beating it be let lie three or four Days, and then let it be beat well over again when 'tis used.

If the Bricks are laid in very hot dry Weather, and it be some small Piece of Work which you would have very strong, it will be worth while to dip every Brick as 'tis laid in a Pan of Water; and it will be of good use in great Work to throw Water on the Walls, after the Bricks are laid, by a *Wirtenburgh* Syphon, or some such way. And the walls also should be covered in Summer Time, to keep them from drying too hastily, as well as in Winter Time from Rain, Snow, or Frost, which last is a great Enemy to all kinds of Mortar, especially to that which hath been newly wetted.

Let the Workmen also in doing up the Walls, not carry any one above three Foot high, before the next adjoining Wall is brought up equal with it, that so they may join together, and make *good Bond* in the Work.

In the Middle of Walls Care should be taken that Joint is not laid on Joint, at least as seldom as may be, that *good Bond* may be made there as well as on the Outsides.

When Timber is laid on Brick-work, as Torsels for Mantle Trees or Lintels for Windows, Templets for Girders, *etc.* let it always be laid in *Loam*, for that is a great Preserver of Timber, which Mortar will eat and Corrode: The same thing should be done with the Ends of Joists and Girders, some Workmen dawb their End with Pitch, to preserve them from the Mortar.

BUTMENTS, in Architecture, are the Mason and Bricklayer's Term for those Supporters or Props, on or against which the Feet of Arches rest. Also little Places taken out of the Yard of the *Ground-Plot* of a House, for a Buttery, Scullery, *etc.* are sometimes called *Butments*.

C

CAMBER *Beam*, in Architecture, is a Beam or Piece of Timber cut hollow or arching in the Middle: They are us'd in Plat-forms, Church Leads, *etc*. and are very proper wherever is occasion for long Beams, being much stronger than flat Beams of the same size; for being laid with the hollow Side downwards, and having good Butments at the Ends, they serve for a Kind of Arch.

CANTALIVERS, in Architecture, are a Kind of *Modilions*, only those are plain, but these are carved. They are much the same with *Cartouzes*, and are set as Modilions are, under the *Corona* of the Cornish of a Buiding.

CARCASE of a House, in Architecture, are the Partitions, Floors, Rafters, *etc*. made by the Carpenter.

CARTONS, in Painting, are the most perfect sort of Drawings on Paper, which are subservient to great Works, that must be thoroughly finished, all Sketches, Drawings, Models, and all other Works being preparatory only, and to be reduced to *Cartons*. This way was practised by *Michael Angelo* and *Raphael Urbin, etc*. the *Cartons* at *Hampton Court* being the Work of *Raphael*.

They are sometimes made by measuring the Height and Breadth of the Place where the Painting is design'd to stand; and then by dividing that *Space* into Squares, and *one* afterwards of the same Dimensions on Paper; they draw on the latter the Subject of their Story or Design, transfering it afterward from the *Carton* to the Wall, by painting it there in Oil, *etc*.

But some Great and Experienc'd Masters will be able to do all this without the help of the *Grate*, as they call it; that is, the Division of the Space or Table into Squares, tho', after all, 'tis a good secure way.

CARTOUZES, in Architecture, are much the same with *Modilions*, only these are set under the Cornish in Wainscotting, and those under the Cornish at the Eves of a House. Some Workmen call them *Dentils* or Teeth.

CASEMENT, the same with *Cavetto*, in Architecture, is a hollow Moulding; some say ¼, some say ⅙ of a Circle in Compass.

CASTING *of Drapery*, is a Term in Painting, signifying a free, easy, genteel, negligent way of cloathing any Figure; and when the Drapery is so managed, they say 'tis *well Cast*.

CAUKING, in Architecture, signifies Dove-tailing across.

CAULICOLI are, in Architecture, the little carved Scrolls which are under the *Abacus* in the Corinthian Capital.

CAVAZION, in Architecture, is the digging or hollowing away of the Earth from the Foundations of a Building. *Palladio* saith it ought to be ¼ of the whole Height of the whole Building.

CAVETTO, in Architecture, is a hollow Moulding of about ⅙ or ¼ of a Circle in Compass.

CEMENT, in Architecture, is a strong binding Mortar. 'Tis used to Cement Bricks together for some kinds of Moulding; to make a *Block of Bricks* for the carving of *Scrolls*, *Capitals*, *etc*. 'Tis of two sorts: The *hot Cement*, which is common, is made by putting to a Pound of Rozin and a Quarter of a Pound of Bees Wax, half an Ounce of fine Brick dust, and as much Powder of Chalk; and boiling all together in a Pipkin, *etc*. for about ¼ of an Hour, stirring it all the while; then let it stand 4 or 5 Minutes, and 'tis fit for use.

The Bricks to be *Cemented* must be heated in or by the Fire, and rub'd one upon another like Boards in glewing. The *cold Cement* is less common, and is thus made.

Grate half a Pound of old Cheshire Cheese, well pared, very small, and to it, in a Pot, put half a Pint of Milk, and so let it stand all Night; the next Morning, put it into a *Tray*, *etc*. With the Cheese and Milk put half a Pound of very fine powdered and sifted quick Lime, and stir all well together with a Trowel, breaking the Knots of the Cheese if any remain; and then add the Whites of 12 or 14 Eggs, which incorporate well with the Mixture, and then it will be fit for use; If you would have it reddish, colour it with a little very fine Powder of Bricks.

CHAPEL, is of two sorts; either *adjoining* to a Church, as a Parcel of the same, which Men of Worth hold, *ut ibidem Familiaria Sepulchra sibi constituant*; or else *Separate* from the Mother Church, where the Parish is wide; and is usually called a *Chapel of Ease*, because Built for the Ease of such Parishioners as dwell too far from the Church. Sometimes this is served by a Curate, provided at the Charge of the Rector, and sometimes at *Theirs* that have Benefit by it, according to Custom or Composition. And some of these latter Chapels which have a Maintenance perpetual by some Lands Charitably bestowed on them, or otherwise, without the Charge of the Rector of the Church; are called *Free Chapels*.

CHAPTRELS, in Architecture, are the same with *Imposts*; and signify those Parts on which the Feet of Arches stand. Their height or thickness ought to be equal to the breadth of the lower part of the Key Stone.

CIMA, in Architecture, is a Moulding somewhat resembling an *S. Vitruvius* makes it of two Quarter Circles joyn'd together. Our Workmen call it an *O.G.*

CIMATUM, in Architecture, is what our Workmen call an *O.G.*, with the Hollow downwards; 'tis part of the Ornament of the Dorick Capital; it stands just above the Square, and hath a *Fillet* over it.

CLAMP NAILS, are such Nails as are used to fasten on Clamps in building and repairing of Ships.

CLAMP'D. When a Piece of Board is fitted with the Grain to the End of another Piece of Board across the Grain, the former Board is said to be *clamp'd*. And thus are the Ends of Tables usually *clamp'd*, to keep them from *warping* or *casting*.

CLARO-OBSCURO, is a Term in Painting; and signifies the Art of distributing Lights and Shadows advantageously, as well on particular Objects as on a Picture in general. On particular Objects 'tis necessary to give them an agreeable Roundness and a convenient Relieve; and in the Picture in general, to expose the Objects with Pleasure to the View of the Spectators, by giving the Eye an Occasion to rest; which is best done by an happy Distribution of great Lights and Shadows, which by their Opposition set off one another.

CLASP-NAILS, are such whose Heads are brought into a little Compass, so that they will sink into the Wood, and when drove home, let a Smoothing Plane go over them.

CLENCH-NAILS, are such as will drive without splitting the Boards, and draw without breaking. They are used by Boat and Barge Builders; and are proper for any Building with Boards that must be taken down again: For fine Work they are made with Clasp-heads.

CLOUT-NAILS, are such as are commonly used for nailing on of *Clouts* (or Plates of Iron) to the Axle Trees of Carriges; and are proper to fasten any Iron to Wood.

COLAPTICE, is the Art of Carving or Cutting the Resemblences and Figures of Natural Things in Stone. The Term for the Artist here is *Lithoxos.*

COLONNADE, in Architecture, is a Range of Pillars running quite round a Building and standing within the Walls of it.

COMMUNICATION. Lines of Communication in the Art of Fortification, are Trenches made to continue and preserve a safe Correspondence between two *Forts* or *Posts*: or at a Seige between two *Approaches*, that they may relieve one another on occasion.

COMPARTITION, in Architecture, is the useful and graceful Distribution of the whole Ground-plot of an Edifice, into Rooms of Office, of Reception or Entertainment.

COMPARTMENT, in Architecture, is a particular Square, or other figured Space (for an Inscription, *etc.*) mark'd out in some ornamental Part of a Building.

COMPOSITION *in Painting*, comprehends two Parts, *Invention* and *Disposition*; one of which finds out (by the means of History) proper Objects for a Picture; and the other places them a-right: the just mixture of which Two Things, according to a true Oeconomy, is properly *Composition*.

CONTRASTE, is a *French* Term used in Painting and Sculpture, and signifies the due placing of the Different Objects and Parts of the Figures; and produces that Variety which is so pleasing in the Attitudes, as if in a *Groupe* of three Figures, one stand out forward, the other backward, and the third appear to the middle between them, they say 'tis well *Contrasted*; So they say also, that to render the Attidudes of Figures agreeable and just, they must be natural and active, varied in their Actions, and *Contrasted* in their Members: So that for Figures to be well Contrasted, is for them to be lively, and to express the Motion proper to the Design of the whole piece, or of any particular *Grouppe*.

CORBETT, in Architecture, is a short piece of Timber placed in a Wall with its end sticking out 6 or 8 Inches as occasion serves. The underpart of the end thus sticking out is sometimes cut into the Form of a *Boultin*, sometimes of an *O.G.*, and sometimes of a *Face*, *etc.* according to the Workman's Fancy, the upper side being plain and flat. These *Corbetts* are usually placed for strengths sake, immediately under the Semigirders of a Platt-form, and sometimes under the Ends of *Camber Beams*; in which latter Case they are commonly placed a Foot or two below the Beam, and have a piece of Timber standing upright close to the Wall, from the *Corbett* to the Beam.

Corbetts, also is a Term used by some Architects, for the hallow Nicks in Walls which are left for Images or Figures, or Statues to stand in.

COUCH, is a Term in Painting, signifying the ground Bed, or Bahs on which any Colour lies; and is distinguish'd from the *Field*, which is always *Horizontal* and upon the *Flat*, whereas the *Couch* may be upright or Vertical, and in any other Position.

COVING-*Cornish*, in Architecture, is such a Cornish that hath a great Casement or Hollow in it, which is commonly Lathed and Plaistred upon Compass sprockets or Brackets; also when Houses are built projecting out over the ground Plott, and then turn'd with a Semi-arch of Timber, which is Lathed and Plaistred; they call that *Coving-Work*.

CRAYONS, are Pencils of several Colours made purposely to paint withal: They are cheifly used in Portraicts; and Pictures done after this way of Painting, are said to be done with *Crayons*.

CROWN-*Post*, is a Post which in some Buildings stands upright in the middle, between Two principal Rafters, and from it there go *Strutts* or *Braces* to the middle of each Rafter. This is also by some called the *King-piece*, and the *Joggle-piece*.

CRYPTOGRAPHY, called also *Steganography*, is the Art of Secret Writing: On which Subject many Books have been publish'd; as, by the Abbot *Trithemius*, in his *Steganographia*, *Francf*. 1608. 4*to*. by *Gustavus Selenus*, in 9 Books in *Fol*. printed at *Luneburgh*, 1624. *cum Fig*. by *Schottus*, in his *Schola Stenographica*. Bp. *Wilkin's Secret and Swift Messenger*.

CULVERTAIL, the same with *Dovetail*; which see.

CUPULO, in Architecture, is an Arched Room or Turret standing on the very Top of a *Dome* or great Building; in Form either Circular or Polygonal; some call it Lanthorn.

D

DESIGN, in Painting or Sculpture, is the Expression of the Images or Ideas that the Painter hath conceiv'd in his Mind, on the Picture, *etc*. and it is Good, when the Author has a good *Gusto* and correct Judgment: This is the Basis and Foundation of all other Parts, and may be compared to the *Stile* of a correct Writer. But the Painters call *Designs* chiefly such Draughts as thay usually express on Paper, in order to the Performance of some considerable Piece of Work: A feint imperfect Design is usually call'd a *Sketch*.

DIAGLYPHICE, is the Art of Cutting, or otherwise making hollow or concave Figures in Metals; such as Seals and *Intaglias's*.

DILAPIDATION, is a wastful destroying, or letting of Buildings run to Ruin and Decay for want of Reparation. 13 *Eliz. c*. 13. And the Money recover'd for Dilapidations, by 14 *Eliz*. 11. shall be employed in the Repair of the same Houses.

DISTANCE *of the Eye*, in Perspective, is a Line drawn from the Foot of the Line of Altitude of the Eye, to the Point where a Line drawn at Right Angles to it will intersect the Object.

DOG-NAILS are such as are used for fastening Hinges.

DORMANT *Tree*, is a Word used in Architecture, by some Workmen, for a great Beam lying a-cross the House; which is usually called a *Summer*.

DORMER, in Architecture, is a Window made in the Roof of a House, and it stands upon the Rafters.

DOUBLE *Aspect*, a Term in Painting. See *Aspect Double*.

DOVETAILING, in Architecture, is a Way of fastening Boards or Timber together, by letting one Piece into another indentedly, with a *Dovetail Joint*, or with a Joynt in the Form of a Dove's Tail.

DRAGON *Beams*, in Architecture, are two strong *Braces* or *Struts* which stand under a Breast Summer, and meet in an Angle on the Shoulder of the King-piece.

E

EASEL-*Pieces*, in Painting, are such small Pieces, either Portraits or Landskips, which are painted on the *Easel*; (the Frame on which the Painter places his strain'd Canvas) and are so called to distinguish them from larger Pictures, which are drawn on Cielings, Roofs, or the Walls of Rooms.

EASEMENT, (*Aisiumentum*) in the Law, is a Service which one Neighbour hath of another, by Charter or Prescription, without Profit; as a Way through his Ground, a Sink, *etc*. and this, in the Civil Law, is called *Servitus Predii*.

EAVESLATH, in Architecture, is a thick feather edg'd Board, nail'd round the Eaves of a House, for the lowermost Tiles, Slate, or Shingles to rest upon.

ENCAUSTICE, is the Art of *Enameling*.

ENCOLAPTICE, is the Art of making Plates of Brass, and cutting in the Figures or Letters for Monumental Inscriptions and Laws.

EXTREMITIES, of the Figures in Painting, are the Head, Hands, and Feet; and these should be drawn with more Nicety and Exactness, or more terminated than other Parts; and must help, by that Means, to render the Action more expressive.

F

FACE *prolong'd*, in Fortification, is that Part of the Line of Defence Razant which lies between the Angle of the Shoulder and the Curtain; or 'tis the Line of Defence Razant diminished by the Length of a Face.

FACIA'S, in Architecture, as the Workmen call it, (for it should be spelt *Fasciæ*,) are nothing but broad Lists or Fillets, and are commonly made in Architraves, and in the Corners of Pedestals. In Brick Buildings, they call the juttings out of the Bricks over the Windows in all Stories but the highest, by this Name; these are sometimes plain and sometimes moulded; and this moulding is only a *Scima Reversa* or an *O.G.*, at the Bottom, and above this are two plain Courses of Bricks, then an Astragal, and last of all, a *Boultine*; which the Workmen call a *Boultrea* and a *Boltell*.

FLUTES, in Architecture, are the Hollows made in the body of a Column or Pillar, and which then is said to be *Fluted*. The Dorick, Ionick, Corinthian and Composite Columns, are usually fluted all along the Body of the Pillar, from the Base to the Capital. Each column hath 24 Flutes, and each Flute is hallowed in exactly a quarter of a Circle. In the Dorick Column the Flutes joyn together without any Interspace, but in the Ionick, Corinthian and Composite Orders, there runs a List between every two Fultes.

FLYERS, is the Workmens Term in Architecture for Stairs that go strait, and don't wind round; nor are the Steps tapering, but the fore and back Part of each Stair and the Ends are respectively Parallel to one another. So that if one Flight don't carry you to your design'd Height, there is a broad half Space; and then you fly again with Steps everywhere of the same Breadth and Length, as before.

FLYING *Bridges*, in Fortification, are made of two small Bridges laid one upon the other, so that the uppermost, by the means of Ropes and Pullies, is forc'd forwards 'till the End of it join to the Place design'd.

FORNICATION, is sometimes used in Architecture to signify Arching or Vaulting.

FRESCO. Painting in *Fresco* is thus perform'd. The Colours are ground with Lime-Water, Milk, or Whey, and temper'd or mix'd together in Pots, as in Size Colouring. The Plaister is made of the Powder of old rubbish Stones, mixt with burnt Flints or Lime, and Water; but the Salt of the Lime must be wash'd out, by often pouring on fresh Water; and to make the Plaister stick the better, they drive in Stumps of Horse-nails about 6 Inches asunder. With this Plaister the Wall is to be laid thick, and then it must be let dry; and afterwards a new Layer of Plaister is put of the Thickness of an Half-Crown; and the Colours must be wrought with a quick free Hand whilst it is wet, for there is no altering of it after it is dry. This Way of Painting was in most ancient Use among the *Greeks*, and from them came to the *Romans*, with whom it was famous. *Raphael Urbin* and *Julio Romano* were eminent this Way; There being three Chambers in the Vatican of their Doing, yet in being.

FURRING, in Architecture, is making good the Rafters Feet in the Cornish: That is, when Rafters are cut with a Knee, these *Furrings*, or *Furrs*, are Pieces that go straight along with the Rafter from the Top of the Knee to the Cornish. Also when Rafters are rotten, or sunk hollow in the Middle, there are Pieces cut thickest in the middle, and tapering towards each End, which are nailed upon them to make them straight; and such Pieces are call'd *Furrs*, and the putting them on, *Furring* the Rafters.

FUST, is a Term in Architecture for the Shaft of a Column, from the *Astragal* to the *Capital.*

G

GABLE-*End* of a House, is the Upright Triangular End, from the Cornish or Eaves, to the Top of its Roof.

GAGE, in Joinery, is an Instrument made to strike a Line truly parallel to the straight Side of any Board or Piece of Stuff. Its chief Use is for gaging of Tenants true to fit into Mortisses, and for gaging Stuff of an equal Thickness. It is made of an Oval Piece of Wood, fitted on upon a square Stick, to slide up and down stifly thereon, and with a Tooth in the End of a Staff, to score or strike a Line upon the Stuff at any Distance, according to the Distance of the Oval from it.

GAIN, in Architecture, is the Workmen's Term for the Bevilling Shoulder of the Joist or other Timber. 'Tis used also for the lapping off the End of the Joist, *etc.* upon a Trimmer or Girder; when the Thickness of the Shoulder is cut into the Trimmer also, and bevilling upwards that it may just receive the *Gain*; and so the Joist and Trimmer lie even and level with their Surface. This Way of Working is used in Floors and Hearths.

GIRDERS, in Architecture, are the largest Pieces of Timber in a Floor: Their Ends are usually fastened into the *Summers* or *Brest-Summers*, and the Joists are framed in at one End to the *Girders*. No *Girder* should lie less than 10 Inches into the Wall, and their Ends should be laid in *Lome*, *etc.*

GLYPHICE, is a Part of Sculpture, being the Art of Cutting, Carving, or Casting, the Images and Resemblances of natural Things in Metals.

GOTHICK *Manner of Building*, was unhappily brought into Use, after the Irruption of those barbarous Nations, the *Goths* and *Vandals*, *etc.* from the *North*, and the *Moors* and *Arabs*, from the *South* and *East*, into the Civiliz'd World. These rude People demolish'd what they could of the Ancient *Greek* and *Roman* Architecture, and instead of those Admirable and Regular Orders and Manner of Building, introduc'd a licentious and fanatical Manner, which, tho' sometimes adorned with Expensive Carvings, and Costly, tho' Lamentable, Imagery, is without any of that August Beauty, and just Symmetry, which the Fabricks of the Ancients entertain us with.

GRAND *Gusto*, is a Term used by Painters, to express, that in a Picture there is something very great and extraordinary, to surprise, please, and instruct. Where this is found, they say the Painter was a Man of the *Grand Gusto*, or *Gout*: And they use the Words *Sublime* and *Marvellous*, when they speak of a Picture, in much the same Sense.

GROTESQUE, or *Grotesca* Work, is the same with what is sometimes call'd *Antick*; being a Confus'd Composure of Figures of different Natures, Sexes, *etc.* and usually of such Fancies as are not any where really existing in *Rerum Natura*.

GROUND *Plates*, in Architecture, are the outermost Pieces of Timber lying on or near the Ground, and framed into one another with Mortesses and Tennons: In these also are Mortesses made to receive the Tennons of the *Joists*, the *Summer*, and *Girders*; and sometimes the *Trimmers* for the Stair-Case and Chimney-Way, and the Binding Joists.

GROUPPE, in Painting or Sculpture, is a Knot or separate Collection of Figures, which appear to have some plain Connexion with, or Relation to one another, by the Design of the Piece.

GUERITE, in Fortification, is a small Tower of Wood or Stone, plac'd usually on the Point of a Bastion, or on the Angles of the Shoulder, to hold a Centinel, who is to take care of the Ditch, and to watch out against Surprizes.

H

HANCES in Architecture are the Ends of Elliptical Arches; and these are the Arks of Smaller Circles than the *Scheme* or Middle part of the Arch.

HATCHING, is a Term us'd in the Art of Drawing or Designing in Black and White; and by it is understood a manner of Shadowing by a continual *Series*, or Succession of many Lines shorter, or longer; closer or more separate; oblique, or direct, according as the Work requires, to render it more or less enlighten'd.

HEADS, is a Term us'd by Builders for those kind of Tiles which they use to lay at the Eaves of a House; being the full breadth of a Common Tile, and but half a Tile in length.

HEIGHT or *Altitude* of the Eye in Perspective, is its *Height* or Elevation above the Horizontal Plane; and it is measured by a Perpendicular let fall from the Eye to that Plane.

HIP-*Roof*, in Architecture, is such a Roof as hath neither *Gable-Heads*, *Shread-Heads*, nor *Jerkin-Heads*: (By which is meant such Heads as are both Gable and Hip at the same End; that is; Gable as far as the Cott or Beam, and then their over short Hips which shut up with their Tops to the Tops of a pair of Rafters, call'd *Singlers* by the Country Workmen.) For a *Hip-Roof* hath Rafters as long, and with the Angles at the Foot, etc. at the Ends of Buildings, as it hath on the Sides: And the Feet of the Rafters on the Ends of such Buildings as have *Hip-Roofs*, stand on the same Plane, viz. parallel to the Horizon, and at the same

Height from the Foundation with the Rafters at the Sides of the Roofs. These *Hip-Roofs* some call *Italian Roofs*.

HIPS, in Architecture, are those Pieces of Timber which are at the Corners of a Roof: They are a good deal longer than the Rafters, because of their oblique Position; for they are level at every Angle. The Country Workmen call them Corners; and by some they are call'd Principal *Rafters*, by others *Sleepers*. As Rafters have four plain Sides, these have usually five.

HYPERTHRE, in the Ancient Architecture, was two Ranks of Pillars all about, and ten at each Face, of any Temple, *etc.* with a Peristile within of 6 Columns.

I

IMPOSTS in Architecture, are what are some times called *Chaptrels*, being the Parts on which the Feet of Arches stand: Or the Capitals of Pilasters that support Arches. These Imposts conform to their proper Orders: The *Tuscan* hath a *Plinth* only: The *Dorick* two *Faces and a Round*: The *Ionick* a *Plancere* or Cavity between the two *Faces*; with, some times, carved Mouldings; as the *Corinthian* and *Composite* have their *Freises*. But the *Sallies* of the *Imposts* must not exceed the Body of the Pilaster. Sometimes the Entablature of the Order serves for the Impost of the Arch; and this looks very great and stately.

INLAND, is an old Word found in *Domesday* and other old Books, and signified that Part of any Land or Mannor which lay next to the Mansion House, and which was used by the Lord himself; whence it was called *Terra Dominicalis*, *Demesnes*, in opposition to *Utland* or *Outland*, which was in Tenancy. These *Inlands* the Feudists call *Terras Curtas*, *ac Intra Curtem*, *Court Lands*, or such as were appropriated to the Court or House of the Lord. So

INLANTAL, *In-land*, or Demesne was opposed to *Delantal*, or *Outland*, or *Land Tenanted*.

INSULATA *Columna*, in Architecture, is a Pillar which stands alone, like an Island, as it were, in the vast Ocean of the Air. *Evelyn's Parallel*.

INTAGLIO'S are Precious Stones engraved with Heads of Great Men, or Inscriptions, *etc.* such as are often set in Rings, or Seals.

J

JOISTS, in Architecture, are such pieces of Timber as are framed into the Girders and Summers, and on which the Boards of the Floor are laid.

JOYSTS: See *Joists.*

K

KEYSTONE in an Arch, is placed at the Top or Vertex of Elliptical or such like flat Arches, to bind the Sweeps of the Arch together.

KING PEICE in any Building, is a Piece of Timber standing upright in the Middle between two principal Rafters, and having Strutts or Braces going from it to the Middle of each Rafter.

L

LAYMAN, among the Painters is a Statue of Wood, whose Joints are so made, that it may be put into any posture; and its chiefest use is for the *Casting* and *Adjusting* of *Draperies* for the cloathing of Figures.

LEAD-NAILS, are such as are commonly used to nail down Lead, Leather or Canvas to hard Wood.

LEDGERS, are long pieces of Timber fastened horizontally to the Poles in the Scaffolds belonging to any Wall or Building, on which the outermost ends of the Putlogs do rest.

LINE *of the Section*, in Perspective, is the Intersection or Contact of the *Plane to be projected* with the *Glass or Diaphanous Plane.*

LOCAL *Colours*, in Painting, are such as are natural and proper for each particular Object in a Picture; and they are so called to distinguish them from the *Claro-Obscuro*, Which see.

LUTHERNS, a sort of Windows in the Roof of a House. See *Dormer.*

M

MAGAZINE or *Arsenal*, is the place in Fortified Towns, *etc.* where all sorts of Stores are kept, and where Carpenters, Wheel-wrights, Smiths, *etc.* are employed in making all things needful to furnish out the Train of Artillery.

MANNER, besides what hath been said about *Manner*, I must take notice that they say in Architecture, That an Order Heroically and Gigantically designed; where the division of the principal Members is put into a few parts; but those having all a bold and ample Relievo, is after the *Grand Manner*. As for example, in a *Corniche*, if the *Gola* or *Cimatium* of the *Corona*; the *Coping*, the *Modillions* or *Dentelli* make a noble appearance by the gracefulness of their Projectures; and that we see none of that ordinary Confusion which results from those little Cavities, *quarter rounds* of the *Astragal*, and such little Ornaments as produce no effect in great and massy Works, and which do impertinently justle out the graceful and principal Members; then will the *manner* of this Corniche appear *solemn* and *great*, and approve it self to be performed after *La Grand Maniere*. In Mr. *Evelyn's* Parallel of Ancient and Modern Architecture, *Chap*. 5. *p*. 25. you have a fine draught of this Grand Manner in an Ancient Dorick Pillar which was found at *Albano* joining to the Church of St. *Mary* near *Rome*.

MANSE, is a Parsonage or Vicarage House for the Incumbent to live in, and was originally, and is now, an essential part of the Endowment of a Parish-Church, together with the Glebe and Tithes.

MASSES, in Painting, are the large parts of a Picture containing the great Lights and Shadows.

MEMBRETTO, in Architecture, is the *Italian* Term for a *Pilaster*, that bears up an Arch. These are often *fluted*, but not with above 7 or 9 Channels. They are frequently used to adorn Door-Cases, Gallery Fronts, and Chimney Pieces, and to bear up the Cornices and Freezes in Wainscot.

MESSUAGE, is a Dwelling-house, with some Land assigned for its use; and by this Name a Garden, Shop, Mill, Chamber or Cellar may pass, saith *Plowden*, *fol.* 169. In *Scotland* it is what we call the *Mannor House*, the principal Dwelling-House within any Barony.

METOPS, the same with *Metopa*: See Vol. I.

MITRE, in Architecture, is the Workmen's Term for an Angle that is just 45 degrees, or half a right one; and if it be a quarter of a Right Angle they call it a *Half Mitre*: And they have an Instrument made to this Angle which they call the *Mitre Square*, with which they strike Mitre Lines on their Quarters or Battens; and for dispatch they have a *Mitre Box*, as they call it, which is made of two pieces of Wood, each about an Inch thick, and one is nailed upright upon the Edge of the other; the upper piece hath the Mitre Lines struck upon it on both sides, and a Kerf to direct the Saw in cutting the Mitre Joints steadily, by only applying the piece into this Box.

MOAT; the Brink of the Moat next the Rampart in any Fortification is called the *Scarp*, and the opposite one the *Counterscarp*.

MODELL, is an original Pattern which any Workman proposes to imitate; 'tis variously made of Wood, Stone, Plaster, *etc*. and should be (in Architecture, for instance) made by a *Scale*, where an Inch or half an Inch represents a Foot, for the more exact compleating of the design.

MORESK-Work or *Morisko-Work*, is a kind of Antick-work in Painting or Sculpture, after the manner of the *Moors*, consisting of several *Grotesco's*.

MORTISE or *Mortaise*, is the Workmens word in Architecture for the Hole made in one piece of Timber to receive the Tennon of another piece.

MOSAIC *Work*, is corruptly so called; it should be *Musaick*; in Latin *Musivum Opus*, and sometimes 'tis called also *Pavimenta Tesselata*.

MUNIONS, in Architecture, are the short upright Posts or Bars that divide the several Lights in a Window Frame.

N

NAVE, in Architecture, signifies the main body of a Church.

NEWEL, in Architecture, is the upright Post that the Case of winding Stairs turns round about.

O

OGEE, *Ogive*, and as it is often written *O.G.*, is a sort of Moulding in Architecture, consisting of a Round and a Hollow. *Vitruvius* makes it two quarter Circles. *Scamozzi* and some others, make the Arches flatter. 'Tis almost in the form of an *S*, and is the same with what *Vitruvius* calls *Cima*. *Cima reversa* is an *O.G.*, with the Hollow downwards, as some define it.

OPENING *of the Trenches*; is the first breaking Ground of the Besiegers, in order to carry on their Attacks against the Town. The difference between *this and carrying on the Trenches* is, that *this* is only the beginning of the Trench: It is begun by a small Foss or Ditch which the Pioneers make in the Night on their Knees; usually about a Musket-shot from the place, but if there be no hollow nor rising Ground to favour them, they begin farther off. This small Foss is afterwards enlarged by the next Pioneers which come behind the first; and so 'tis dug deeper by degrees till it be about 12 Foot broad and 5 Foot deep. The Earth that is dug out is thrown up as they go along, and serves for a Parapet to save them from the Fire of the Town. The place where the Trenches are open'd is called *the End of the Trench.*

ORDONANCE, in Painting or Sculpture, is the just and elegant Composition of the whole Piece by a proper, natural and agreeable disposition of the Figures, so as to answer the Design of the History.

OVOLO or *Echinus*, in Architecture, is a part of the Ornaments or Mouldings of the Cornish of a Pillar; which in the Tuscan and Dorick Orders is turned like a *Scima* or *Cymatium*, and is substituted for the support of the *Corona*; in the *Dorick Order* it usually hath a slender *Regula* above it, and in the *Corinthian* both above and below too, where it is likewise carved and adorned with a broad *Welt* like a *Plinth.*

P

PAINTING. Books treating of this Art and of the Eminent Artists are as followeth.

An Idea of the Perfection of Painting: Originally written in French by Rowland Treart, *sieur* de Cambray, *and rendred English by* J. Evelyn, *Esq*; *F.R.S.* Lond. 1668. 8*vo.*

A General Idea of the Art of Painting, and a Relation of 7 Conferences held at Paris in the Academy Royal for the improvements of the Arts of Painting and Sculpture.

Optique de Portraiture et Peinture, contenant la Perspective Speculaire et Practique Accomplic. etc. *per Gregoire Huret de l'Academ. Royale de Peinture et Sculpture, à Paris.* 1670. *Fol.*

Vol. II.

Entretiens sur les vies et sur les Ouvrages des plus Excellens Peintres Anciens et Modern per M. Felibien.

PALLIFICATION, in Architecture, is the *Pileing* the Ground-Work, or strengthening it with Piles or Timber driven into the Ground, when they build upon a *Moist* or *Marshy Soil. Builder's Dictionary.*

PALLISADES *turning*, are an Invention of Mr. *Coehornes*, for in order to preserve the Pallisades of the Parapet from the Besiegers Shot; he orders them so, that as many of them as stand in the length of a Rod, or in about 10 Foot, turn up and down like a Trap; so that they are not in sight of the Enemy, but only just when they bring on their Attack; and yet are always ready to do the proper service of Pallisades.

PARADIGRAMMATICE, is the Art of making all sort of Figures in Plaister. The Artists in this are called *Gypsochi.*

PARASTÆ, in Architecture, are the same with *Pilasters*; the *Italians* call them *Membretti*.

PARGETING or *Parging*, is the Workmens word for Plaistering of Walls.

PEDIMENT, in Architecture, is a Triangular Frame with Cornish raised over the Front of a House, and sometimes over Doors, *etc*.

PEERS, in Architecture, are a kind of Pilasters or Buttresses for Support, Strength, and Sometimes Ornament.

PIAZZA'S, or as our Vulgar frequently call them *Piaches*, are in the *Italian* the same as our Cloysters.

PIEDROIT, in Architecture, is a Square Pillar which is partly within the Wall. *Builder's Dictionary*.

PILÆ, in Architecture, and their *Quandra's* and Tables (as we yet see them in ancient Altars and Monuments) were imploy'd for Inscriptions; but if shorter and more massy, they serve for Arches of Bridges and for Buttresses to solid Work. *Evelyn's Parallel.*

PITCH, is a word used by Architects and Builders in these Senses. Sometime *Paving* is called *Pitching*: But usually they understand by it the Angle which a Gable-end, and consequently the whole Roof of a Building is set to. If the Length of each *Rafter* be ¾ of the Breadth of the Building, then they say that Roof is of a *True Pitch*: But if the Rafters are longer, they say 'tis a *high* or *sharp pitch'd Roof*; if shorter, they call it a *low* or *flat pitch'd Roof.*

PLANE, is an Instrument used in Joinery to make Boards Plane, Try and Smooth, or in order to joint or frame them together, *etc.* These are of several Names and Sorts according to the several Uses; as, 1. The *Fore-Plane*, which is a long Plane, and used first of all before either *Smooth-Plane* or *Jointer*. The Edge of the Iron of this Plane is not ground streight, but rising with a Convex Arch in the middle of it, that its Edge may bear to be set the *Ranker*, for its use is to take off the Irregularities of the Stuff as soon as may be, in order to prepare it for the *Smoothing-Plane* or *Jointer*. If the Stuff be *free* and *frowy*; that is, even tempered all over, you may then set the Plane so *rank* as that you may take off a Shaving of the thickness of an old Shilling; but if it be hard or curling, you can't take off one thicker than an old Groat. 2. The *Smoothing-Plane*, is a short small Plane, whose Iron is set very fine, and its use is to take off the Irregularities of the *Fore-Plane*, or those which it hath left. 3. The *Jointer* is a Plane longer than the *Fore-Plane*, and hath its *Sole* perfectly strait from End to End; it comes after the *Fore-Plane* and *Smoothing-Plane*, and is design'd to shoot an Edge of a Board perfectly streight in order to jointing, as also Boards of any thickness; for 'tis used to *Try*, as they call it, that is, smooth Tables with, whether large or small: wherefore its Iron must be set very fine, little above an Hair's breadth above the Sole of the Plane, and the length of the Edge of it exactly strait, or parallel to the Plain of the Sole or Bottom of the Plane. 4. The *Strike Block*, is a Plane made as true as the *Jointer*, and like it, only shorter; being used also to shoot a short joint, which it doth more handily than the *long Jointer*. 'Tis used also for framing and fitting the Joints of *Mitres* and *Bevels*, but then the piece of Wood is drawn by hand over the Plane several times till 'tis shot true. 5. The *Rabbet Plane*, is used to cut part of the upper Edge of a Board strait or square down into the Stuff, so that the Edge of another Board cut down after the same manner may fit and join in with it on the Square; and when two Boards are thus cut away this lapping over is called *Rabbeting*. The *Rabbet-Plane* is sometimes used also to strike a *Fascia* in a piece of Moulding. The Iron of this Plane is full as broad as the Stock is thick (usually about an Inch) that the Angles of the Edge may cut down exactly strait; and it delivers its Shavings at the Side and not out of a Mouth at the Top, like other Planes. 6. The *Plow*, is a narrow *Rabbet-Plane*, with the addition of two Staves with Shoulders to them, and on the Bottom of the Shoulders a Fence: Its use is to plow a narrow square *Groove* on the Edge of any Board, of any proper depth. 7. *Moulding-Planes*, of which are several kinds, as the Round Plane, the Hollow, the O.G., the Snipes Bill, *etc.* and these of several Sizes, as from half an Inch to an Inch and half. When these Planes are used on soft Wood, as Deal, Pear-Tree, Maple, *etc.* the Iron is set to an Angle of 45 Degr. with the Sole or Base of the Plane; but if it be very hard Wood, as Box, Ebony, *Lignum Vitæ*, *etc.* it is set to 80 Degr. and sometimes quite upright. There is also some difference in the grinding of the *Basil* or the Slope of the Edge of the Iron of the Plane; for in Working on hard Wood this is ground to an Angle of about 18 or 20 Degrees; but in soft Wood, not to one above 12 Degrees; for the more acute the *Basil* is the better and smoother the Iron cuts, but the more obtuse and thicker it is, the stronger is the Edge to cut upon hard Work.

PLASTICE, or the *Plastick Art*, is the Art of making Figures of Men, Birds, Beasts, Fishes, Plants, *etc.* in Clay, *etc.* The Workmen are called *Plastæ*. It differs from Carving, because here the Figures are made by *Addition* usually, but in Carving always by *Subtraction* of what is superfluous. It is now with us chiefly used in Fretwork Ceilings; but the *Italians* apply it to the Mantlings of Chimneys with great Figures.

PORTRAITS, is the Painters word for Pictures of Men and Women (either Heads, or greater Lengths) drawn from the Life; and the word is used to distinguish this kind of Face-Painting (as it is often called) from History-Painting.

PRICK-*Posts*, in a Building, are such as are framed into the *Breast-Summers* between the Principal Posts for strengthening the Carcase of the House.

PRINCIPAL *Posts*, in any wooden Building, are the Corner Posts, which are tennanted into the Ground Plates below and the Raising Plates above, *i.e.* into the Beams of the Roof.

PRINTS. The Original of *Prints* or *Cutts*, as we sometimes call them, was this; in the Year 1460, one *Maso Finiguerra*, a Goldsmith of *Florence*, graved his Plate; and then casting some of it in melted Brimstone, he perceived, that what came out of the Mold was markt with the same Prints as his Plate, by the Black which the Sulphur had taken from the Graving: after this he tried to do as much on Silver Plates with wet paper, by rolling it Smoothly with a Roller, and this Succeeded. This Novely tempted *Baccio Baldini*, a Goldsmith of the same City, to attempt the same thing, which he did with Success; ingraving Several Plates of *Sandro Boticello's* Invention, and Design; and upon this *Andrew Mantegna*, who was then at Rome, Set about engraving some of his own Pieces. This knowledge getting into *Flanders*, *Martin* of *Antwerp*, a famous Painter, graved abundance of Plates of his own Invention, and sent several Prints into *Italy*, which were markt thus, *M. C.* After him the Famous *Albert Durer* appear'd and gave the word a vast number of Prints, both in wood, and copper. About this time one *Hugo de Carpi* an *Italian* Painter of no great Capacity, but of a ready Invention, found out a way, by means of several Plates of wood, to make Prints resemble Designs of *Claro-Obscuro*: and some Years after the Invention of *Etching* was discovered, which was soon made use of by *Parmeggiano*.

PROFILE, in Fortification, is the Representation of the Height, Depth and Thickness of any Work.

PROSTYLE, in the ancient Greek Architecture, was a Rank or Row of 4 Columns only, and whose Station was in the Front of a Temple, or other great Building.

PULVINATA, in Architecture, is the Term for a *Freeze*, which swells out like a Pillow.

PUNCHINS, in Architecture, are short pieces of Timber placed to support some considerable Weight: They commonly stand upright between the Posts, and are shorter and slighter than either the *Principal Posts* or *Prick-Posts*. Those that stand either side of a Door are called *Door Punchins*.

PURLINS, in Architecture, are those Pieces of Timber which lie across the Rafters on the Inside, to keep them from sinking in the middle of their Length.

PUTLOGS, are short pieces of Timber (about 7 Foot long) used in Building Scaffolds; they lie at Right-Angles to the Wall with one of their Ends resting upon the *Ledgers* or Poles which lie parallel to the side of the Wall of the Building.

Q

QUARTERS, (in Architecture) are those slight upright Pieces of Timber which are placed between the Punchions and Posts, they are used to *Lath* upon.

QUIRK, in a Term in Architecture for a Piece of Ground whether Square or Oblong, taken out of a Corner, or any Place else of a Ground-plat, to make a Court or Yard, *etc*.

QUOINS are the Stones and Bricks placed in the Corner of any Building; and if they stick without the Brick-work (their Edges being Cypher'd off) they are called *Rastick Quoins*.

R

RAFTERS, in any Building, are those Pieces of Timber, which stand by pairs upon the *Reason*, meet in an Angle at the Top, and help to compose the Roof of a Building: They should not stand farther than 12 Inches from one another.

REFECTORY was that Place in a Monastery where the Monks, Friars, Nuns, *etc*. usually Dined and Supped.

REPOSE, is a Term in Painting, signifying the Place where the *Masses*, or great Lights and Shadows are assembled: And this being well understood hinders the Confusion of Objects; suffering not the view to be contrasted altogether, but to proceed gradually and successively without Disturbance.

RIGLET, is any square, flat, thin, Piece of Wood, like those which are designed to make the Frames of small Pictures of; which are so called, before they are *Molded*.

S

SAPHETA, in Architecture, is the Board over the top of a Window, and placed Parallel, and opposite to the Window board at the Bottom.

SCENOGRAPHICK *Projection*, or Perspective, is the Transcription of any given Magnitude, into a Plane which intersects the Optick Pyramid at a proper distance: For in Projection, there is to be consider'd, 1. The *Object*, or Foundation, or Ground of the Projection, from whence the *Pyramid*, *Cone*, or *Pencil of Rays* go. 2. The *Eye* of the Spectator, and 3. The *Plane Table*, or *Diaphanum*, which intersects the Rays, some where between the Object and the Eye: And the Representation or Appearance of the Object in that *Plane*, is the *Projection* or *Perspective* of the Object. This *Plane* is always supposed to be at Right Angles with the Horizon. And from hence it will follow, that (1.) A *Point* will be projected there in the Diaphanous Plane where the Optick Ray cuts it. (2.) That Right Line will be projected where the Optick Triangle and the

Plane do mutually Intersect each other. (3.) That a *Plane* or Superficies will be represented where the *Diaphanous Plane* cuts the Optick Pyramid of Rays coming from that Surface. And that Representaion of it called its *Image*.

SCHEAME, or *Skeen*, in Architecture, is the Workmens word for the middle part of an Elliptical Arch.

SCRATCH-*Work*, in *Italian*, *Sgrafitti*, was a way of Painting in *Fresco*, by preparing of a black Ground, on which was placed a white Plaster; and this White being taken off with an Iron Bodkin, the Black appears thro' the Holes, and serves for Shadows: This kind of Work is lasting, but being very rough, is unpleasant to the sight. 'Twas used in *Rome* by *Polidoro de Caravagio*.

SCRIBING, when the Joyners would fit a piece of Board, *etc*. to an irregular Surface, or any other irregular Piece: They open their stiff Iron Compasses to the greatest distance, any where between the 2 Boards, *etc*. and then carrying one Leg along all the irregular Indentings, *etc*. of one, the other Leg moving Parallel to it, describes that irregular Figure on the other Board; which being in that manner cut will fit and joyn.

SCULPTURE, may be distinguished into three several Arts, each of which hath its Specifick Difference. For (saith Mr. *Evelyn* in his History of *Chalcography*) besides *Sculptura*, as it relates to *Chalcography*, there is both *Sculptura*, and *Cœlatura*; both which, according to *Quintilian*, differ from the *first*, with respect to the *Matter*, on or out of which any thing is wrought. For it was applied to cutting or carving in *Wood* or *Ivory*; and then was called *Tomice*, and the Artists *Desectores*; to working in Plaster, and then called *Paradigrammatice*, and the Artists, *Gypsochi*; to Cutting or Carving in Stone, and then called *Colaptice*, and the Workmen *Lithoxoi*, and lastly in Metals, *Glyphice*. And it may be described *to be an Art, which teaches us to cut or take away all that is superfluous of the subject Matter, reducing it to that Form or Body, which was designed in the Mind of the Artist.*

SELL, in Architecture, is the Term both for the lowest piece of Timber in a Timber-building, or for that on which the whole Superstructure is erected; and also for the bottom-pieces in a Window-frame; the former is called a *Ground-sell*, the latter a *Window-sell*.

SOFITTO, is the *Italian* Term in Architecture, for the *Eaves* of the *Corona* of the Capital of a Column.

STILES, the upright pieces which go from the bottom to the top in any Wainscot, are by the Workmen called *Stiles*.

STILOBATUM, is in Architecture the Body of the Pedestal of any Column.

STRIÆ, are the Lists or Rays which run between the *Flutes*, *Chanels* or *Striges* in Fluted Pillars. These are 20 in the Dorick, and 24 in the Ionick Order.

STRIGES, in Architecture, are the hollow Chanels in the Shaft of a Column, called by our Workmen *Flutings* and *Grooves*. They more properly belong to the Ionick Order, tho' they are found sometimes in the Dorick, and often in the Corinthian and Composite. They are generally at right Angles with the Plane of the Base, but are sometimes found winding about the Pillar. Sometimes the *Striges* are filled up with a Swelling a third part from the Base, lying in the hollow like a Stick shot of a round form: These Mr. *Evelyn* thinks we should call *Staved* or *Cabled Columns*.

STRUT, is a Term used by some Builders for that *Brace* which is framed into the *King-piece* and the *principal Rafters*.

STUCCO, Painting in *Stucco* was revived from the Ancients by *Giovanni d'Udine* a Scholar of *Raphael's*; he found the true Matter which the Ancients made use of, which was a Composition of Lime and Marble Powder'd very fine.

T

TAILLE *douce*, a Term in Painting, signifying, as Mr. *Evelyn* in Chalcography tells us, the Art of Sculpture or Chalcography it self. In *French* it signifies *sweetly* or *tenderly cut*, and this, whether done with *Burin* (or Graver) or with *Aquafortis*, which we call *Etching*. He saith the *Italians* call it *Intaglia* or *Stamp*, without any Epithet or Adjunct; and also *Bolino*, which he takes to be the more ancient and unwarrantable, as implying the use of the *Point* and *Needle*, and of *Etching* with *Aquafortis*; which is sometimes so happily performed, as scarce to be discerned from the finest Stoaks of the *Bolio*, or *Graver* it self.

TALUS *Superior*, in Fortification, is the Slope on the top of the Parapet; for the Top of the Parapet is made sloping, that the Soldiers may defend the Covert-way with small Shot, which they could not do if it were level.

TASSELS, in Building, are those pieces of Boards that lie under the Ends of the Mantle-trees.

TENAILLE *of the Foss* or Ditch, is a low Work raised before the Curtain in the middle of the Foss, it is of three sorts. The first is Composed of a Curtain, two Flanks and two Faces. The *Rampart* of the *Curtain*, including the *Parapet* and *Talus*, is but five Fathom thick, but the Rampart of the *Flanks* and *Faces* is seven. The Second, which *Vauban* saith, he found to be of very good Defence, is Composed only of two Faces made on the Lines of Defence, whose *Rampart* and *Faces* are Parallel. The Third sort differs from the Second, only that its Rampart is Parallel to the Curtain of the Place. All three sorts are good Defences for the Ditch, and lie so low, that they cannot be hurt by the Besieger's Cannon, till they are Masters of the Covert-way, and have planted their Cannon there.

TENON, in Architecture, is the square end of a piece of Timber fitted into a Mortise.

TESSELATA *Pavimenta*, were the Pavements in the Tents of the Roman Generals, a rich *Mosaic Work*, made of curious small square Marbles, Bricks, or *Tiles*, called *Tessellæ*, from the Form of *Dies*.

THORUS, see Tore [Volume I].

TIGE, in Architecture, is the Shaft of a Column from the Astragal to the Capital.

TOMICE, the Art of Carving in Wood or Ivory. The Artist in this Work was called *Desecter*.

TRANSOM, in Architecture, is the piece of Timber which is framed a-cross in a double light Window.

TRAVERSE, in Fortification, is a Trench with a small Parapet, and sometimes with two, one on each side; it serves as a Cover from the Enemy when they come on their Flank. Sometimes 'tis covered over with Planks on the top, and is also loaded with Earth; they are of good use to stop an Enemy's Way, and to prevent being Enfiladed. It is also a good Defence in a dry Ditch, when the Parapet is made on that side next the opposite Flank. There is also a *Traverse* in a wet Ditch, which is made by throwing into the Foss or Ditch over against the Place, where the Miner is to be put to the Foot of the Wall, Saucissons, Joysts, and other pieces of Wood, with Fascines, Stones, Earth, *etc.* in order to fill up the Ditch, and to carry a Gallery over it. Also a Wall of Stone or Earth, built a-cross a Work which is commanded, in order to cover the Men, is called a *Traverse*.

TRIMMERS, in Architecture, are those pieces of Timber Framed at Right Angles to the Joists against the ways for Chimneys, and Well-holes for Stairs. *Builder's Dictionary*.

V

VIVO, is the Shaft of a Column in any of the Orders of Pillars in Architecture.

W

WATER-*Table*, in Architecture, is a sort of Ledge left in Stone or Brick-Walls, about 18 or 20 Inches from the Ground, and there the thickness of the Wall begins to abate.

WINDING-STAIRS, are such as wind-round a Newel, either Circular or Square, Close, *i.e. Solid*, or Open.

Z

ZACCO, is the Term sometimes used for the lower part of the Pedestal of a Column; and is in the Form of a Square Brick or Tile.

ZOPERUS, a Term in Architecture, signifying the same with *Frieze*. See Vol. I, *Frize*.

References

The following references are the primary sources consulted for biographical and historical information and also for the quotations and citations to be found in the various texts. Given the nature of the interrelationships between the several volumes in our series *The Encyclopaedic Dictionary in the Eighteenth Century*: *Architecture, Arts and Crafts*, we have grouped together, for the readers interest and convenience, all the references consulted in the preparation of the individual volumes listed under the short title of each volume.

John Harris: *Lexicon Technicum*

Chapter 1

1. These ideas are outlined in the preface to Bacon's *The New Organon* for which see:
   R. L. Ellis and J. S. Spedding, Edited by John M. Robertson, *The Philosophical Works of Francis Bacon, Baron of Verulam, Viscount of St. Albans, and Lord High Chancellor of England*, George Routledge and Sons Ltd., New York: E. P. Dutton & Co., 1905.
2. The essay *Of Building* is adapted from the text given in note 1.
3. The essay *Of Gardens* is adapted from the text given in note 1.
4. The reader interested in Sir Francis Bacon's water gardens will find a description of these in:
   Paula Henderson, *Sir Francis Bacon's Water Gardens At Gorhambury*, The Journal of the Garden History Society, Vol. 20, Number 2, 1992, pp. 116-131.

Chapter 2

1. The biographical account of the life of Sir Henry Wotton is adapted from:
   Thomas Zouch, *The Lives of Dr. John Donne*; *Sir Henry Wotton*; *Mr. Richard Hooker*; *Mr. George Herbert*; *and Dr. Robert Sanderson*, *by Isaac Walton*. Volume I, York: Payne and Foss, and Wilson and Sons, 1817.

Chapter 4

1. Thompson Cooper, *John Harris*, The Dictionary of National Biography, Vol. IX, London: Oxford University Press, 1959-60 (Reprinted).
2. John Harris, *Lexicon Technicum: or, an Universal English Dictionary of ARTS and SCIENCES: Explaining not only The TERMS of ART, But The ARTS Themselves.* Volume I, London: MDCCIV.
3. For a discussion of eighteenth-century encyclopaedias, and in particular that of John Harris, see:
   Lael Ely Bradshaw, *John Harris's Lexicon Technicum*, Voltaire Studies, no. 194, 1981: pp. 107-121.
4. Sir Henry Wotton, *The Elements of Architecture*, John Bill, London: 1642.
5. As we have noted in the text, the second volume of *Lexicon Technicum* was published as a 'Supplement' to the first as:
   John Harris, *Lexicon Technicum: or an UNIVERSAL English Dictionary of ARTS and SCIENCES: Explaining not only The TERMS of ART, But The ARTS Themselves.* Volume II, London: MDCCX.

Ephraim Chambers: *Cyclopaedia*: *Volume I*

Chapter 1

1. Francis Espinasse, *Ephraim Chambers*, The Dictionary of National Biography, Vol. IV, London: Oxford University Press, 1908.
2. Gerald Le Grys Norgate, *John Senex*, The Dictionary of National Biography, Vol. XVIII, London: Oxford University Press, 1963-64.
3. Ephraim Chambers, *Prospectus*; *Proposals for Printing by Subscription*, *Cyclopaedia*, London, January 1710.
4. For a discussion of eighteenth-century encyclopaedias, and in particular that of Ephraim Chambers, see:
   Lael Ely Bradshaw, *Ephraim Chambers' Cyclopaedia*, Voltaire Studies, no. 194, 1981: pp. 123-140
5. For a discussion of the work of Ephraim Chambers in its wider context see:
   Robert Collison, *Encyclopaedias: their history throughout the ages*, New York: 1964.
6. The Rev. Alexander Gordon, *Abraham Rees*, The Dictionary of National Biography, Vol. XVI, London: Oxford University Press, 1909.
7. Ephraim Chambers, *CYCLOPAEDIA: or, an UNIVERSAL DICTIONARY of ARTS and SCIENCES.* Two volumes, London: MDCCXXVIII.

8. For a discussion of the transition of Chambers' *Cyclopaedia* into Diderot's *Encyclopédie* see:
Terence M. Russell, *Architecture in the Encyclopédie of Diderot and d'Alembert: The Letterpress Articles and Selected Engravings*, Aldershot: Scolar Press, Ashgate Publishing Ltd., 1993.
9. My understanding of the writings of Francis Bacon has been derived through consultation of the original texts (see Chapter 1, note 1 of John Harris) and:
F. H. Anderson, *The Philosophy of Francis Bacon*, Chicago, Illinois: The University of Chicago Press, 1948.
10.For some discussion of the endeavours to translate Chambers' *Cyclopaedia* into French see:
J. G. Alger, *John Mills*, The Dictionary of National Biography, Vol. XIII, London: Oxford University Press, 1909.

The Builder's Dictionary

Chapter 1

1. For a discussion of *The Builder's Dictionary* in the context of eighteenth-century architectural books and such precursors as Joseph Moxon's *Mechanick Exercises* see:
Eileen Harris assisted by Nicholas Savage, *British Architectural Books and Writers 1556-1785*, Cambridge: Cambridge University Press, 1990.
2. As we remark in the text, our specific reference here is to the work of Francis Bacon concerning his *Preparative towards a natural and Experimental History*. For the complete reference, see Chapter 1, note 1 of John Harris.
3. For Robert Ellis's introduction to the works of Sir Francis Bacon see Chapter 1, note 1 of John Harris.
4. The formal library-catalogue reference for the first edition is:
The Builder's Dictionary: Or, Gentleman and Architect's COMPANION. Explaining not only the TERMS of ART In all the several PARTS of ARCHITECTURE, But also containing the THEORY and PRACTICE Of the Various Branches thereof, requisite to be known by Masons, Carpenters, Joiners, Bricklayers, Plaisterers, Painters, Glaziers, Smiths, Turners, Carvers, Statuaries, Plumbers, etc. London: Printed for A. Bettesworth and C. Hitch, at the Red-Lion in Pater-noster-Row; and S. Austen, at the Angel and Bible in St. Paul's Church-Yard. MDCCXXXIV.
For the reader unable to locate a copy of the first edition, a facsimile is available in the following form:

The Builder's Dictionary: Or, Gentleman and Architect's Companion. Being A Complete UNABRIDGED REPRINT of the earlier work published by A. Bettesworth and C. Hitch. Washington: Association For Preservation Technology, 2 Vols., MCMLXXXI.

Samuel Johnson: *A Dictionary of the English Language*

Chapter 1

1. Quotations throughout Chapters 1 and 2, unless otherwise indicated, are taken from the opening chapters of Boswell's Life of Johnson in the following edition:
   James Boswell, *Life of Johnson*, London: H. Ingram & Co., Vol. I., 1851.
2. My understanding of Johnson has been greatly assisted by my reading of Mr. Wain's scholarly and moving account:
   John Wain, *Samuel Johnson*, London: MacMillan Ltd., 1974, reprinted 1980.
3. For insights into the everyday objects in Johnson's life and for portrayals of his contemporaries, I am indebted to Ms. Lane's richly illustrated study:
   Margaret Lane, *Samuel Johnson and his World*, London: Hamish Hamilton Ltd., 1975.

Chapter 2

1. Information relating to Johnson and his London circle has, in part, been derived from:
   Leslie Stephen, *Samuel Johnson*, The Dictionary of National Biography, Vol. X, London: Oxford University Press, 1963-64 (Reprinted).
2. For an overview of Johnson at the time of his work on the *Dictionary*, I have benefited from consulting:
   Sir Sydney Castle Roberts, *Samuel Johnson*, Chicago and London: Encyclopaedia Britannica, 1972.
3. Samuel Johnson, *The Plan of a Dictionary of the English Language*, London: 1747.
4. My understanding of many matters bearing upon the creation of Johnson's *Dictionary* has been aided by the views of:
   J. D. Fleeman, *The Genesis of Johnson's Dictionary*, London: Longman, 1990 (?).
5. A description of Johnson's working method based upon the most recent research is given by:
   Allen Reddick, *The Making of Johnson's Dictionary 1746-1773*. Cambridge: Cambridge University Press, 1990.

6. Robert Harrison, *Richard Bathurst*, The Dictionary of National Biography, Vol. I, London: Oxford University Press, 1963-64 (Reprinted).
7. Cecil J. L. Price, *Philip Dormer Stanhope, Fourth Earl of Chesterfield*, Chicago and London: *Encyclopaedia Britannica*, 1972.

Chapter 3

1. The tables have been compiled from, and the letterpress articles have been extracted from, the first edition of the *Dictionary*:
   Samuel Johnson, *A DICTIONARY OF THE ENGLISH LANGUAGE: In Which The WORDS are deduced from their ORIGINALS, and ILLUSTRATED in their DIFFERENT SIGNIFICATIONS by EXAMPLES from the best WRITERS.* London: Printed by W. Strahan, For J. and P. Knapton [etc.], 2 Vols., MDCCLV.
2. For an assessment of Johnson's accomplishment placed in a wider context see:
   Brian O'kill, *The Lexicographic Achievement of Johnson*, London: Longman, 1990 (?).

Encyclopaedia Britannica

Chapter 1

1. For an outline of the lives of Bell, Macfarquhar and Smellie and their contribution to the creation and publication of the first edition of Encyclopaedia Britannica I have benefited from the following:
   John Armitage, *Andrew Bell, Colin Macfarquhar and William Smellie*, Chicago and London: Encyclopaedia Britannica, 1972.
2. Bernard Barham Woodward, *William Smellie*, The Dictionary of National Biography, Vol. XVIII, London: Oxford University Press, 1959-60 (Reprinted).
3. *Encyclopaedia Britannica, Or, A Dictionary Of Arts and Sciences, Compiled Upon A New Plan In Which The different Sciences and Arts are digested as they occur in the order of the Alphabet*, Edinburgh: Printed for A. Bell and C. Macfarquhar . . . at his Printing Office, Nicolson-Street. MDCCLXXI.

Additional Works by the same Author

As stated in the Preface, the present work *The Encyclopaedic Dictionary in the Eighteenth Century: Architecture, Arts and Crafts* has its origins in previous bibliographical researches undertaken by the author. For the reader interested in these matters, the following publication details are provided:

Architecture in the Encyclopedié of Diderot and d'Alembert: The Letterpress Articles and selected Engravings. Scolar Press, Aldershot 1993. pp. xii, 224. 100 Plates and Frontispiece. This book traces the origins and publication history of the *Encyclopédie* and provides a detailed inventory, of 5,000 architectural writings abstracted from the original 17 folio volumes of text and 11 folio volumes of plates.

The Built Environment: A Subject Index 1800-1960. 4 Volumes. Gregg Publishing, London 1988. The books in this index were researched over a ten-year period from the extensive British Library Catalogue, with the approval of the copyright authorities and received substantial government financial support. The index gives detailed citations to more than 70,000 printed books across the whole realm of the built environment.

Volume 1:	Town Planning and Urbanism Architecture Gardens and Landscape Design	1120 pp
Volume 2:	Environmental Technology Constructional Engineering Building and Materials	960 pp
Volume 3:	Decorative Art and Industrial Design International Exhibitions and Collections Recreational and Performing Arts	1152 pp
Volume 4:	Public Health Municipal Services Community Welfare	1232 pp

Bibliographical Index: A Subject Guide to Printed Books in the Fields of Architecture, Building, Constructional Engineering, Environment, Technology, Planning, Landscape Architecture, Design and allied Arts and Crafts. University Microfilms International (UMI) Ann Arbor, Michigan USA 1982. This work was compiled over a ten-year period from the British Library Catalogue with the consent of the copyright authorities. Each volume has an introduction in French, German, Italian, and Spanish. The 32 volumes which constitute the index are each substantial monographs which are published (in the USA) in both microfiche and conventional hard-back book form. The titles in the series are listed below together with the number of pages for each volume. The number of bibliographical entries contained within each volume is shown in brackets: each citation is listed under a subject heading with details of the author, book title and all relevant publication details.

	Pages	Citations
ARCHITECTURE		
Vol. 1 Architectural history and theory	178	(2579)
Vol. 2 Domestic architecture	115	(1322)
Vol. 3 Ecclesiastical and sacred architecture	170	(2162)
Vol. 4 Egyptian, Greek and Roman antiquities	128	(1660)
Vol. 5 National architecture	177	(2397)
STUDIES OF BUILDING TYPES		
Vol. 1 Cinema art and craft	117	(1561)
Vol. 2 Factories and workshops	100	(1214)
Vol. 3 Hospitals and hospital work	107	(1209)
Vol. 4 Libraries	200	(2643)
Vol. 5 Prisons	94	(1106)
Vol. 6 Theatres and the theatre	223	(3165)
Vol. 7 Miscellaneous buildings	217	(2949)
BUILDING AND CONSTRUCTIONAL ENGINEERING		
Vol. 1 Architectural engineering	152	(1911)
Vol. 2 Building	280	(3555)
Vol. 3 Structure and materials	288	(3571)
PLANNING, HOUSING AND PUBLIC HEALTH		
Vol. 1 Hygiene and welfare	261	(3644)
Vol. 2 Public health	262	(3615)
Vol. 3 The home and household management	92	(1206)
Vol. 4 Town planning and urbanism	179	(2168)
Vol. 5 Working class housing	223	(2168)
ENVIRONMENT AND TECHNOLOGY		
Vol. 1 Acoustics, lighting and colour	170	(2017)
Vol. 2 Electricity and electrical engineering	218	(2849)
Vol. 3 Heating, ventilation and gas engineering	214	(2754)
Vol. 4 Irrigation, drainage and sanitation	161	(2019)
Vol. 5 Water and water supply	171	(2092)
DESIGN, FURNITURE AND DECORATION		
Vol. 1 Aesthetics	85	(1171)
Vol. 2 Design, decoration and industrial art	218	(2977)
Vol. 3 Furniture and interior design	174	(2038)
Vol. 4 International and industrial exhibitions	113	(981)
Vol. 5 Museums, art galleries and collections	228	(3112)
GARDENS AND LANDSCAPE ARCHITECTURE		
Vol. 1 Gardens and gardening	283	(3934)
Vol. 2 Landscape architecture, botanic parks and gardens	141	(1711)

Index

The Explanation of the Plate for the Five Orders of Architecture.

In the Tuscan *Column.*

a The *Ovolo* or *Egg*: In the *Ionick* and *Corinthian* Orders, and every where by some, 'tis called the *Echinus*. The *French* call it the *Quart de rond*.
b The *Astragal*, or *Baguette*, as the *French* call it: The *Italians* the *Tondino* or *Round*; below which is a small List, Reglet, or Fillet.
c Is the *Crown*, the *French* call it the *Larmier*.
e The *List* or *Fillet* under that.
f The *Gula* or *Talon* in *French*.
g The *List* of the *Gula*.
H The *Frieze*.
i The List of the *Architrave*.
k The *Architrave*.
l The List of the *Abacus*.
m The *Abacus Cymatium* or *Saillon*, as the *French* call it.
n The *Ovolo* or *Echinus*.
o The *List*, *Ring*, or *Fillet* under that.
p The Frieze of the Capital: The *French* call it, The *Gorgerin* or *Collarin*.
q The *Astragal* of
r The *Congè* or *Ceinture*, as the *French* call it; or of the Neck of the Column.
s The *Body* of the Column; in *French* the *Fust* or *Vif*.
t The *Listel* of the Base, or the *Netherband* or *Swath*.
u The *Tore*, or the *Gros Baton*, in *French*.
w The *Plinth*.
x The *Reglet*, *List* or *Fillet*.
y The *Gula* reversed, or the *Talon*.
z The *Ring*, *List* or *Fillet*.
B The *Die* of the Base.
2 The *Reglet* or *Fillet*.
3 The Base of the *Pedestal*; in *French*, the *Zocle*.

In the Dorick *Column.*

a The *Gula*, sometimes the Grove or hollowing of the upper List.
b The *Dentils* or Teeth.
c The *Capital* of the *Tryglyph*.
d The *Tryglyph*, of which those Parts that are framed Inwards or hollowed, are called *Flutes*: And the Square of the Frieze between the *Tryglyph*, is called the *Metops*, as *r*; and sometimes the outermost towards the Right-hand is called a *Demi-Metops*.
e The *Cymatium* or *Bandelette*.
f The *Guttæ* or *Drops*.
g The *Platteband* or *Face*.
h The *Reglet*.
i The *Talon* or *Gula*.
k The *Ovolo* or *Echinus*.
l Three *Annulets* or *Fillets*.
m The *Frieze*, *Gorgerin* or *Collarin*.
S The *Fust*, *Shaft* or *Body* of the Pillar.
o The *Listel*.
p The *Tore*.
q The *Plinth*.
r The *Fillets* and *Gula*.
B The *Die* of the Base.
t The *Fillets* or *Reglets* of
u The Base of the *Pedestal*.

In the Ionick *Column, the Things in which it differs from the former, are these.*

1. The *First Scotia.*
2. The *Second Scotia.*
3. The *Bands* or *Canal* of the *Volutæ.*
4. The *Volutæ.*
5. The First or *Little Face.*
6. The Second or *Middle Face.*
7. The Third or the *Great Face.*
9. The Figure of *Eggs* carved there.

In the Corinthian *Column, the Things peculiar are,*

1. *A Tore, Reglet* and *Gula*, finely wrought.
2. *A Reglet* and *Astragal*, with
3. *Frieze* above it.

4, 5, 6. The Ranks or Rows of Leaves, some say of the *Acanthus* or *Great Dock*, others of *Olive*, and others of *Palm.*

8. The *Abacus* of the *Capital.*
9. The *Flower.*

In the Composite *Order.*

The *Pedestal* of this *Column*, is the same with that of the *Corinthian*, except in the Members of the *Cymatium* and *Base.*

And the *Composite Capital* differs from the *Corinthian* only in this, That it hath *Volutæ* or *Scrolls* like the *Ionick*, when the *Corinthian* hath curled Stalks; and therefore 'tis called *Composite*, because composed of the *Ionick* and *Corinthian* together.

TUSCAN DORICK IONICK CORINTHIAN COMPOSITE

Plate: *Architecture*.

The five orders of architecture, namely: Tuscan, Dorick, Ionick, Corinthian, and Composite – see text.

The Explanation of the Plate for Fortification.

A *A Single Bastion.*

B *A Double Bastion* or *Cavalier*, to overlook the Enemies Bastions, and to scour their Trenches.

C *A Platt-Bastion*; which is made when the Line of Defence is too long; *i. e.* exceeds 150 Fathom, which is the Distance a Musket will be sure to do Execution. Wherefore this Platt-Bastion is placed in the middle of the Curtain.

D *A Half-Bastion* placed on the Side of a River.

E *A Bastion Tenailled*; when the Angle of the Bastion is less than 70 Degrees.

F *A Ravelin*, which covers the Gates, Bridges and Curtains. Sometimes these Ravelins have Flanks made to them as G.

G *A Ravelin* with Flanks.

H *An Half-Moon*, made to cover the Flanked Angle of a Bastion. The *Gorges* of *Half-Moons* are circular.

I *Counter-Guards*, which are raised instead of *Half Moons*.

K *A Single Tenaille.*

L *A Queue D'aronde* or *Swallow's-Tail*: So called, because its Sides instead of being parallel, open or spread towards the Head, and grow narrow at the *Gorge*.

M *A Bonnet a Prestre*, or *Priest's Cap*: Some call them *Double Tenailles*. But they differ from a *Tenaille*, because its Sides are not parallel, but spread or open out towards the Campaign like a *Swallow's-Tail*. This Work hath two Angles *Saillant*, and two Inwards.

N *A Horn-work* whose Head is fortified with two Demi-Bastions, or *Epaulments*, joyn'd by a Curtain, and closed by parallel Sides ending at the *Gorge* of the Work. These *Horn-Works* are used to cover the Gates, *etc.* instead of *Tenailles*.

N 2, *A Horn-work*, whose Sides are not parallel.

O *A Crown-Work*, whose Design is to cover some large Spot of Ground to defend a Rising-Ground, or to defend the Head of a Camp that is intrenched: 'Tis the largest of all Out-works, having a large *Gorge*, and two Sides, ending towards the Campaign in two Demi-Bastions; each of which joins by a Curtain, to an entire Bastion that is at the Head of the Work.

P *Tenailles* to defend the Ditch.

Q *An Half-Moon* covered with two *Counter-Guards* R R, to make it the stronger.

S S *Places of Arms* on the *Counterscarp*, being open Spots of Ground for the Garrison to Rendezous on an Alarm, *etc.*

↗ The *Ditch.*

T T, *etc.* The *Counterscarp*, or Covered-way.

g g, etc. The *Glacis.*

V *A Retrenchment* within or behind a Breach; as X.

W *A Place* for a *Magazine.*

l l l Pallisades.
a a Redoubts.
b b b Batteries.
c Lines of Communication.
d d Trenches, by which the Approaches are carried on.
e A Mine under the Glacis and Covered-way of the *Horn-work.*
f A Lodgment at the Foot of the *Glacis.*

Explanation of the Profile.

r a	The Base of the *Rampart*	70 Feet.
t s, q o	Its Height	16
r s	The Inward *Talu*	16
o a	The Outward *Talu*	8
l q	The Base of the *Parapet*	21
Z *d*	The Inward Height	6
n p	The Outward Height	4
n q	The Outward *Talu* of the *Parapet*	2
o d	The Inward	1
y l	The Breadth of the *Foot-Bank*	3
y x	The Height of it	1½
t y	The *Terre-Plane*	25
a e	The *Berme*	3
e g	The Breadth of the Moat	112
i f	Its Depth	12
f h	Its Breadth at the Bottom	88
e s k g	Two *Talu's*, each	12
c g	The little Ditch at the Bottom of the Moat	18
q w	Its Depth	5
g c	The *Covert-way*	18
c d	The Seat of the *Parapet*	60
c f	Its Height	6
f d	The *Glacis.*	

0 50 100
Scale of Feet for the Profile

Plate: *Fortification*.

A representation of the principal Ramparts, Parapets, Moats, and Bulwarks – see text.